ATLA Monograph Series
edited by Dr. Kenneth E. Rowe

1. Ronald L. Grimes. *The Divine Imagination: William Blake's Major Prophetic Visions.* 1972.
2. George D. Kelsey. *Social Ethics Among Southern Baptists, 1917–1969.* 1973.
3. Hilda Adam Kring. *The Harmonists: A Folk-Cultural Approach.* 1973.
4. J. Steven O'Malley. *Pilgrimage of Faith: The Legacy of the Otterbeins.* 1973.
5. Charles Edwin Jones. *Perfectionist Persuasion: The Holiness Movement and American Methodism. 1867–1936.* 1974.
6. Donald E. Byrne, Jr. *No Foot of Land: Folklore of American Methodist Itinerants.* 1975.
7. Milton C. Sernett. *Black Religion and American Evangelicalism: White Protestants, Plantation Missions, and the Flowering of Negro Christianity, 1787–1865.* 1975.
8. Eva Fleischner. *Judaism in German Christian Theology Since 1945: Christianity and Israel Considered in Terms of Mission.* 1975.
9. Walter James Lowe. *Mystery & The Unconscious: A Study in the Thought of Paul Ricoeur.* 1977.
10. Norris Magnuson. *Salvation in the Slums: Evangelical Social Work, 1865–1920.* 1977.
11. William Sherman Minor. *Creativity in Henry Nelson Wieman.* 1977.
12. Thomas Virgil Peterson. *Ham and Japheth: The Mythic World of Whites in the Antebellum South.* 1978.
13. Randall K. Burkett. *Garveyism as a Religious Movement: The Institutionalization of a Black Civil Religion.* 1978.
14. Roger G. Betsworth. *The Radical Movement of the 1960's.* 1980.
15. Alice Cowan Cochran. *Miners, Merchants, and Missionaries: The Roles of Missionaries and Pioneer Churches in the Colorado Gold Rush and Its Aftermath, 1858–1870.* 1980.
16. Irene Lawrence. *Linguistics and Theology: The Significance of Noam Chomsky for Theological Construction.* 1980.
17. Richard E. Williams. *Called and Chosen: The Story of Mother Rebecca Jackson and the Philadelphia Shakers.* 1981.
18. Arthur C. Repp, Sr. *Luther's Catechism Comes to America: Theological Effects on the Issues of the Small Catechism Prepared In or For America Prior to 1850.* 1982.
19. Lewis V. Baldwin. *"Invisible" Strands in African Methodism.* 1983.
20. David W. Gill. *The Word of God in the Ethics of Jacques Ellul.* 1984.
21. Robert Booth Fowler. *Religion and Politics in America.* 1985.
22. Page Putnam Miller. *A Claim to New Roles.* 1985.
23. C. Howard Smith. *Scandinavian Hymnody from the Reformation to the Present.* 1987.
24. Bernard T. Adeney. *Just War, Political Realism, and Faith.* 1988.
25. Paul Wesley Chilcote. *John Wesley and the Women Preachers of Early Methodism.* 1991.
26. Samuel J. Rogal. *A General Introduction of Hymnody and Congregational Song.* 1991.
27. Howard A. Barnes. *Horace Bushnell and the Virtuous Republic.* 1991.
28. Sondra A. O'Neale. *Jupiter Hammon and the Biblical Beginnings of African-American Literature.* 1993.
29. Kathleen P. Deignan. *Christ Spirit: The Eschatology of Shaker Christianity.* 1992.
30. D. Elwood Dunn. *A History of the Episcopal Church in Liberia, 1821–1980.* 1992.
31. Terrance L. Tiessen. *Irenaeus on the Salvation of the Unevangelized.* 1993.
32. James E. McGoldrick. *Baptist Successionism: A Crucial Question in Baptist History.* 1994.

33. Murray A. Rubinstein. *The Origins of the Anglo-American Missionary Enterprise in China, 1807–1840.* 1995.
34. Thomas M. Tanner. *What Ministers Know: A Qualitative Study of Pastors as Information Professionals.* 1994.
35. Jack A. Johnson-Hill. *I-Sight: The World of Rastafari: An Interpretive Sociological Account of Rastafarian Ethics.* 1995.
36. Richard James Severson. *Time, Death, and Eternity: Reflecting on Augustine's* Confessions *in Light of Heidegger's* Being and Time. 1995.
37. Robert F. Scholz. *Press toward the Mark: History of the United Lutheran Synod of New York and New England, 1830–1930.* 1995.

TIME, DEATH, AND ETERNITY
Reflecting on Augustine's *Confessions* in Light of Heidegger's *Being and Time*

by
Richard James Severson

ATLA Monograph Series, No. 36

American Theological Library Association
and
The Scarecrow Press, Inc.
Lanham, Md., & London

SCARECROW PRESS, INC.

Published in the United States of America
by Scarecrow Press, Inc.
4720 Boston Way, Lanham, Maryland 20706

4 Pleydell Gardens, Folkestone
Kent CT20 2DN, England

This book is based on the author's dissertation, "Time, Death, and Eternity: Reflecting on Augustine's *Confessions* in Light of Heidegger's *Being and Time*," University of Iowa, December, 1990

British Cataloging in Publication Information Available

Library of Congress Cataloging-in-Publication Data

Severson, Richard James, 1955–
Time, death, and eternity : reflecting on Augustine's Confessions in light of Heidegger's Being and time / Richard James Severson.
  p.   cm. — (ATLA monograph series : no. 36)
Based on the author's thesis, University of Iowa, 1990.
Includes bibliographical references and index.
1. Augustine, Saint, Bishop of Hippo. Confessiones. 2. Heidegger, Martin, 1889–1976. Sein und Zeit. 3. Ricoeur, Paul. Temps et récit.
4. Time. 5. Death. 6. Eternity. 7. Philosophical theology.
I. Title. II. Series.
BR65.A62S48   1995   236'.1'09—dc20   95–7904

ISBN 0–8108–3012–4 (cloth : alk. paper)

Printed in the United States of America

Contents

Editor's Foreword

Since 1972 the American Theological Library Association has undertaken responsibility for a modest monograph series in the field of religious studies. Our aim in this series is to publish two studies of quality each year. Titles are selected from studies in a wide range of religious and theological disciplines. We are pleased to publish Richard J. Severson's *Time, Death, and Eternity,* a study of Augustine's *Confessions.* Following undergraduate studies at South Dakota State University, Richard Severson studied library science and completed the doctorate in theology and ethics at the University of Iowa. He has held teaching posts at the University of Iowa and at the University of Northern Iowa. Severson has served as reference librarian at several colleges in Iowa and Oregon before assuming that post at Marylhurst College, Marylhurst, Oregon in 1995. The author of several articles, this is Dr. Severson's first book.

Kenneth E. Rowe
Series Editor
Drew University Library
Madison, NJ 07940
USA

Introduction

In the land of death you try to find a happy life: it is not
there. How can life be happy where there is no life at all?
                                Confessions, Book IV, ch. 12, p. 82

Every generation has had a mortality rate of 100 percent.
Human beings have always had to face that fact. Today, however,
we are confronting some new wrinkles to the age-old problem of
death. We live under the threat of nuclear annihilation.[1] We are
dealing with bioethical questions that constantly challenge traditional life categories. Machines can force terminally ill, unconscious persons to breathe and pump blood, keeping them "alive"
in some radically unordinary sense. New technological marvels in
medicine force us to rethink what death itself means.[2] The marvels
of medicine have also changed the sociology of our dying. We
used to die at home, among family and friends. Now we are rushed
by ambulance to hospitals where medical specialists fight for our
lives (to the point of absurdity sometimes, though always well-
intentioned) and where we end up dying alone, often in a
drug-induced state of unawareness.[3]

There can be no doubt that technological advances in medicine,
warfare, transportation, and so forth, have forced us to reshape—
both consciously and unconsciously—our attitudes toward death
and dying. But there is another important, and probably not
unrelated, dimension to our new situation relative to this age-old
problem. It concerns the history of philosophy and theology. I
shall call it the problem of "ontological death." The central
purpose of this work is to depict the meaning and significance of
ontological death and to suggest a theological way to overcome its
threat. Let me briefly articulate the context in which this purpose
is carried out before elaborating the central terms of the argument.

By "ontological death" I specifically have in mind Heidegger's
notion of *Sein zum Tode* as developed in his *Being and Time.*[4]

1

According to Heidegger, death is the one unsurpassable possibility for the self or Dasein. Only from the standpoint of Dasein's
anticipated death can Dasein grasp the meaning of its own being
and thereby achieve authenticity. Death becomes ontological for
Heidegger because Dasein's being is defined by its "dying," that
is, by Dasein's mode of being toward death. Heidegger claims that
death is the final horizon within which to understand the meaning
of human being in the world. Heidegger's sense of ontological
death is adopted for purposes of critical analysis in this study.
Against Heidegger, however, I shall argue that it is still possible to
think of ontological death as open to eternity. The argument
proceeds on theological grounds.

In developing a theological response to Heidegger's assertions
concerning ontological death, this work can be placed within a
delimited tradition of thought within contemporary theology.
Specifically, I draw from the broad post-Kantian stream of
Continental theology, especially as it has been shaped by Paul
Tillich, Rudolph Bultmann, and, more recently, Robert Scharlemann. For this tradition, the Kantian arguments against metaphysical forms of theology are convincing. "Postcritical theology," as
this tradition is sometimes called, posits an idea of God that
attempts to move beyond the Kantian criticism while it at the
same time accepts it.

According to Kant, the object of (metaphysical) theology is the
idea of "the thing which contains the highest condition of the
possibility of all that can be thought."[5] God, conceived as the
highest being or being as such, is a necessary and universal idea of
pure reason. The system of reason logically presupposes the idea
of "God" for its systematic completeness. However, through the
exercise of pure reason alone we cannot know whether God exists
in reality, for knowledge entails the synthesis of a concept with
intuitions exhibitable in sense experience. The idea of God cannot
be exhibited in sense experience; whatever is so exhibited can
present only a conditioned reality and not the unconditioned being
implied by the idea of God. Even if the theistic tradition of
metaphysical thought makes the assertion that God is the highest
being whose properties of unity, eternality, goodness, and so on,
are aspects of the essence of God, it cannot include the claim that
"God exists." In response to the Kantian criticism of metaphysical
theology, two successor concepts arose.[6]

The first of these asserts that "God" names the ultimate agent behind the many activities we observe in the world: God is the one in the many who appears indirectly in the many, just as the one source of light appears indirectly in the many things that are illuminated. Humans do not "know" the one in the many but, rather, "discern" it. In this sense of the term "God," the task of theology is that of "penetrating the particular agents and actions so as to see and to show the one agent which is be-ing in all events" (Scharlemann, *The Being of God,* p. 68). To discern the agency of God is to move from a set of agents and actions to an underlying subject, just as in the process of understanding a game, according to Hans-Georg Gadamer, one's attention is drawn from the playing of the game by several people to the underlying subjectivity of the game itself.[7]

The problem with this version of postcritical discernment of God is that it eludes satisfactory verification. Even if it is possible for theological language ("In these events, God is") to disclose the one in the many such that the word "God" seems to bring out the universal activity of the underlying subject within a manifold of historical activities, there is no criterion to determine why these events and not others should play this role. Just because one person discerns the agency of God in the fall of the Berlin Wall does not mean that another person will understand the event likewise. Two people may agree that "God" names the one subject underlying many activities, yet fail to agree about which activities disclose the ultimate subject. A second person may see the fall of the Berlin Wall as a manifestation of the irrepressibility of human ambition and not the being of God. The idea of God as one in the many provides no criterion for discriminating between the two cases.

The second postcritical idea of God focuses not on the universal form of God's manifestation in worldly events but on the specific content of a divine manifestation. Here we approach the notion of the concrete "symbol" as the self-manifestation of God. More than any other theologian, Tillich has developed this form of theological reflection. This study's theological response to the Heideggerian claim of the ontological primacy of being toward death proceeds within the general context of this idea of God as self-manifesting in the worldly symbol.

The idea of the symbol of God is a more subtle and dialectically sophisticated idea than the idea of God as the one in the many. The dialectic involved can be illustrated by reference to the differences between Kant and Hegel on the ontological proof for the existence of God. From Kant's refutation of the ontological proof for the existence of God, we gain the thought that what is meant by the idea of God cannot manifest itself in existence. From Hegel's reinstatement of the ontological proof, we gain the thought that what is meant by the idea of God cannot do otherwise than manifest itself in existence. The idea of God as self-manifesting in the symbol of God holds both of these thoughts together: God is God in the activity of manifesting divine being in the nondivine as symbol of God; yet the symbol of God is not literally God. Let me focus on the two sides of this dialectic.

Kant argued convincingly that existence is not a predicate (*Critique of Pure Reason*, A592/B620); hence, one cannot derive the existence of God from the concept of God as perfect being. Kant used the idea of a currency to illustrate his point: "Whether I imagine the mere possibility of a hundred Taler or the reality of a hundred Taler, my financial condition remains the same." God is, for Kant, an idea of the ultimate unity of subjectivity and objectivity, which is necessary for the completeness of reason. The difficulty is that the idea of God is the idea of what *cannot* appear to intuition. The idea of God as highest being is the idea of something that cannot be thought to appear in human experience without the contradiction of turning the unconditioned into something conditioned. That than which none greater can be conceived cannot be that than which none greater can be conceived if it can be conceived as something existent.

Hegel, however, in his *Lectures on the Philosophy of Religion*, reaffirmed the substance of the ontological proof, in opposition to both Anselm and Kant.[8] According to Hegel, Anselm's formulation of the idea of God as the perfect being does not express the substance to which his thinking is aiming. The idea of God is, for Hegel, the idea of the manifest openness between idea and reality; it is not a question of perfections. Moreover, Anselm begins with what should be the result of the proof: The definition of God as that than which none greater can be conceived is first presupposed, and then Anselm concludes to God's necessary existence. But the proof should properly conclude with the definition of God, not begin with it in the first premise of the argument.

Hegel likewise criticizes Kant. Anselm was on the right track, whereas Kant made a basic blunder. In likening the idea of God to the idea of a hundred taler, Kant confuses a genuine idea (*Begriff*) with a mere mental picture. Kant's objection holds for the "representation" of a hundred taler but not for the idea of God. The idea of God is unique because it can point to the reality of God in such a way that makes it impossible to separate it from the reality it means. The idea of God, or absolute spirit, is not like an image or picture of a conditioned and finite entity.

Hegel approaches the idea of God by way of finite reality: One recognizes the finite is finite with respect to the conception of God as infinite. But the idea of God as infinite is not the opposite of the finite. The opposite of the finite is not the infinite but another aspect of the finite. The finite subject is not opposed to the infinite subject; instead, it is opposed to the finite object. The infinite appears as the reconciliation of the opposition in finitude itself (between subject and object). Whereas both Anselm and Kant assume that the basic opposition is between human and God, finite and infinite, Hegel claims such an opposition posits a "bad infinite." The infinite cannot be infinite if it is opposed to the finite. The finite subject negates the finite object (or vice versa) and the "true infinite" is the negation of that negation.

When applied to God, the pattern of double negation works as follows: If we form the subjective idea of God, as did Anselm and Kant, we do in fact see that the idea is not the reality itself. The subjective idea is opposed by the objective reality of God, which is found in the religious representation of God. But we can recognize that the objective reality is not what is meant by the idea. The opposition between subjective idea and objective reality is taken up (*aufgehoben*) by means of the thought of God as infinite idea that dynamically comprises both the subjective idea and the objective reality. God as absolute spirit comprises both the abstract concept and the representation of the appearing God in a phenomenological history of religion. God is the dynamic activity through which the idea manifests itself in the reality of religious representations and cancels that manifestation.

Admittedly, Hegel's "proof" is less than clear. In his discussion of the ontological proof, he does not say *where* or *as what* the self-manifesting God is supposed to appear. But his intention to establish God as self-manifesting idea that must appear somewhere *is* quite clear.

The details of Kant's refutation and Hegel's reaffirmation are not themselves at stake in this discussion. Rather, I intend to point out the intellectual background informing Tillich's notion of the religious symbol as the manifestation of the idea of a self-manifesting God. Tillich's thought of the being of God as the religious symbol draws on both Kant and Hegel. We misread Tillich's thought on the religious symbol as the self-manifesting and self-negating presence of God unless we read it against the double background of Kant and Hegel. The religious symbol for Tillich is not merely a figural representation of the literally unrepresentable idea of God as "being itself." The religious symbol is, for Tillich, the self-manifestation of God as being itself. The agency behind the production of the symbol of God is twofold: Both a human agency and the divine agency are at play.

In Robert Scharlemann's Tillichian terminology, the religious symbol is the "being of God when God is not being God." By this, Scharlemann means that the idea of God is the paradoxical idea of being itself, which both cannot and must appear in reality. Roughly following Kant, Scharlemann claims that God is the idea of the final unity of subjectivity and objectivity for which no intuitions can be displayed. And roughly following Hegel, Scharlemann claims that God is at the same time the idea of the final unity of subjectivity and objectivity that, when properly conceived, must manifest itself in the finite realm of the "extra-divine" in order to be the idea of the "true" divine and the "good" infinite. In this sense, the religious symbol is the "otherness" of the idea of God, the divine manifestation in the region of the nondivine. The idea of God must include both the infinite substance of divinity and the finite appearance of divinity as the nondivine. This dynamic appears in Tillich's analysis of the characteristics of the religious symbol and the criteria for the truth of the religious symbol.

In *Dynamics of Faith,* Tillich defines the religious symbol in terms of its ability to point beyond itself while participating in that to which it points.[9] The religious symbol expresses an ultimate concern that transcends the finite realm of human being and discourse, and it also breaks into the finite realm. It is quite clear from Tillich's discussion of the symbol as the condensation of the Word of God that he means to view the religious symbol not only as a human construct but also as a divine self-manifestation.[10]

This is especially evident in Tillich's discussion of the criteria for the truth of the religious symbol. According to Tillich, a "true" religious symbol must meet both subjective and objective criteria of truth (*Dynamics of Faith,* 95–98).

Subjectively considered, the symbol must have the capacity to bind together subject and object by imparting ultimate concern through the symbol. For Tillich, this tests whether the symbol genuinely represents God: The symbol presents to experience the meaning of the idea of God if it enables the overcoming of the subject-object split by binding the subject with the object in concern for the ultimate. Does the symbol evoke ultimate questions about the meaning of being? Does it "grasp" one through its power of calling to mind in an existentially compelling way the idea of God? Clearly, when Tillich discusses religious experience through symbol as an experience of "being grasped by an ultimate concern," he ascribes a quasi agency to the symbol and, through it, to God.

Objectively considered, the symbol must have the capacity to distinguish itself as real symbol from the divine being. In other words, the symbol must have the capacity to call attention to the fact that it is not God. Tillich refers to this as the self-negating power of the true religious symbol. A true symbol both manifests God and denies that in any literal sense it *is* God. Through this self-negating capacity, the symbol distinguishes between the divine being and itself as the otherness of divine being. By insisting that the symbol makes God present and perceptible and yet is not God, the symbol symbolizes the merely symbolic nature of all representations of God. The true religious symbol thus resists idolatry and protects the double nature of God as both infinite idea, unpresentable in time and space, and finite manifestation, the embodiment of divine otherness. Once again, in speaking about the symbol's power of self-negation, Tillich ascribes a quasi agency to the symbol and, through it, to God. The religious symbol is the self-manifestation of God, but never in an exhaustive sense of manifesting God finally or without remainder.

Following Tillich and Scharlemann, this study argues that Augustine's *Confessions,* like other classic religious texts, is still able to present the eternal in a way that transcends our existential world wherein having a care is defined by an ultimate relation to death. Although Augustine may not convince us of the boundedness of time and space by eternity through metaphysical arguments, the text of the *Confessions* can function like a religious symbol to present the being of God when God is not being God; that is, to present eternity as the existential overcoming of ontological death viewed as the ultimate finitude of the human subject.

In *Being and Time,* Heidegger develops an interpretation of Dasein—the human self as the "there" of being in its disclosure—in terms of temporality. His claim is that the wholeness and authenticity of Dasein's temporal nature can only be established through a primordial relationship to death. Dasein can achieve full authenticity or self-understanding only when it heroically lives within the constraints of an awareness that human destiny is determined by death. In Stoic fashion, Dasein's own conscience compels Dasein to anticipate death and to bring death to bear on every aspect of its existence. Thus, death is the limiting horizon for Dasein in its temporal being, which means that *Sein zum Tode* constitutes Dasein's most basic mode of being.

Heidegger's interpretation of the self stands in stark contrast to a traditional interpretation. Dasein is understood in terms of temporality. Yet the traditional view is that the self includes an immortal soul. Heidegger's view of temporality as the meaning of Dasein—particularly when it is associated with human finitude—and immortality are incompatible concepts. There is very little trace of an immortal soul in Heidegger's thought. The transcendent reality that is threatened by Heidegger's *Sein zum Tode* is extratemporal. Clearly, the call of conscience in Heidegger's thought is self-transcending: Authentic Dasein calls to inauthentic Dasein from a futural perspective that is defined by Dasein's most real possibility, which is death. Only when anticipating its ultimate end or death (which involves projection into the future, i.e., self-transcendence) can Dasein assure itself that it is not a victim of self-deception. But the "transcendence" in this case is always confined within Dasein's own temporality. The future, as authentic possibility, transcends the present moment of inauthenticity. In what follows, Heidegger's view of Dasein's call of conscience is contrasted to Augustine's analysis of the immortal soul as it is called to "authenticity" by the Eternal *Verbum* of God. Here the call comes from beyond Dasein's temporality, or from beyond Dasein's own consciousness. Our ability to appreciate this sort of transcendence is threatened by the problem of ontological death, which refers to the increased philosophical (and, likewise, theological) significance of death in the modern world.[11]

If we follow Heidegger's influential position, then death achieving an ontological status means it has become the ultimate horizon for human self-understanding, or for human being in the world. Ontological death seemingly tolls a great human loss: the loss of religious transcendence (where the divine, as extratemporal reality, confronts the human in its temporality) and the loss of transcendent possibilities for human self-identity.[12] As the possibility of a metaphysical realm has decreased in our best philosophical judgment, the philosophical significance of death has increased. Even though the religious significance of eternity in its relation to time is not *necessarily* tied to metaphysical interpretation, it is so tied *practically* by long-standing tradition. Thus, the development of metaphysical thought often determines the development of religious and theological thought. That is why Heidegger's discussion of death is potentially threatening to religious meaning. *Being and Time* actually continues the global criticism of metaphysics that Kant began in the *Critique of Pure Reason.* In fact, there is a line of interpretive progression that runs from Kant to Heidegger.[13]

In the *Critique of Pure Reason,* Kant exposed humans' native tendency toward transcendental illusion.[14] In order to coordinate our concepts and judgments, Kant argued, we must formulate ideas that extend beyond the realm of possible knowledge. In the past, we always assumed that these regulative ideas—of the self as free, the world as a spatiotemporal totality, and God as unconditioned or absolute—constituted real, knowable entities. But according to Kant, such metaphysical assumptions take reason too far. We cannot know anything that is not subject to sensible intuition. Neither the soul nor the cosmos nor God can be intuited. Hence, it is beyond our limited abilities to know whether or not there is such a thing as a soul and whether or not God exists. The "noumenal" realm of things in themselves, or reality as such, is beyond our ken. We must rein in our metaphysical tendencies and "expose the illusions of a reason that forgets its limits" (*Critique of Pure Reason,* A735/B763). Whatever is not subject to the temporal and spatial forms of intuition must remain unknowable.

Within Kant's new, chastened stance toward metaphysics, time especially takes on a heightened significance. He claims that time, as the pure form of inner intuition, is the condition of the possibility for any judgments we might make concerning the objective world. Perhaps an analogy between Kant's regulative ideas of reason and Plato's forms can illustrate the increased importance of time in Kant's philosophy.

The ideas of reason actually serve a similar function in Kant's philosophy to that of the forms in Plato's earlier, "precritical" philosophy. But Plato's forms are transcendent, metaphysical entities to which the mind supposedly has access. With the help of these forms or paradigmatic ideas, so the theory goes, we are enabled to recognize and judge things in the world.[15] The suspicion is that unless we have some familiarity ahead of time with what we are dealing (in this case, the objective world), we will feel alienated. Knowledge presupposes familiarity. The truly alien is neither recognizable nor knowable. Kant subscribes to this truism as much as Plato did.[16] But Kant denies that we have access to a transcendent, metaphysical realm of "prefamiliarity" that enables us to anticipate the spatiotemporal world that we experience in life. The theory of forms takes a necessity of reason too far. The theory of forms argues that we "must think" the form of the highest good, therefore the highest good exists in reality outside of thinking. Kant denies that this conclusion follows. He shows that it is not necessary to think that a necessity of thought provides adequate warrant for claiming the existence of a nonsensible reality. Kant gives us a different, more modest account of our "prefamiliarity" or anticipation of the knowable world.

For Kant, all knowledge involves the synthesis of concepts and intuitions. Since concepts and intuitions are different in kind, some third thing that has similarities to both must be the basis for their union. That third thing is time. On a purely formal or transcendental level, it is through the imaginative variations of temporal constructs—half concept, half intuition—that knowledge can be achieved. For Kant, contrary to Plato, our prefamiliarity with the world of knowable experience is given when we imaginatively/temporally vary the categories or rules of understanding such that they can conform to (and anticipate the possibility of) a manifold of empirical intuitions. The transcendental time determinations of the productive imagination delimit the extent of our possible experience; they anticipate experience, giving us a sense of familiarity with the world. The ideas of reason, on the other hand, only play a regulative role in Kant's account of our experience and knowledge. These ideas are not the basis for our knowledge (as in Plato's forms); instead, they help direct it toward an unconditioned, universal horizon. Again, the ability to anticipate our experience in the knowable world depends upon the work of the transcendental imagination, which focuses on the synthetic qualities of time.

In the context of Kant's critical philosophy, time displaced the role that metaphysical entities such as Plato's forms could play in epistemology. Heidegger's *Being and Time* expresses that displacement more definitively. His title alone is telling: Time is put on a level with being! In the old, precritical (Platonic) metaphysics, being was always associated with an eternal realm that transcends temporality and the finitude that necessarily attends it.[17] With Heidegger's philosophy, however, we seem to have reached the point where time displaces eternity and denies its transcendent claims. As time attains a kind of new equivalence with being, death is granted a new ontological status. Human being in a world deprived of a transcendent, eternal realm is determined by *Sein zum Tode.*

Ontological death, or death in its new role as the exclusive horizon against which we must test our courage and achieve authenticity, symbolizes the death of metaphysics and the consequent loss of familiar transcendent categories. Ontological death threatens the traditional sense of eternity as a never changing presence. It threatens the possibility of religious and theological meaning since eternity is, traditionally, one of God's attributes. *Sein zum Tode,* it seems, even threatens the hope that enabled St. Paul to boast that death cannot be victorious.[18] The sting of death, in its ontological form, seems overpowering. What can we do about this new situation?[19] Must we live our lives without recourse to what lies beyond death? Is it still possible to speak of the presence of eternity within time after the status of time has increased to the point where death—time's consequence—achieves ultimacy?[20] These are questions with which this study wrestles and attempts to answer.

Two responses to this dilemma present themselves. First, one can attempt to defend traditional metaphysical categories by defending theistic arguments and arguments for the immortality of the soul against Kantian and post-Kantian criticisms of these arguments. Many serious efforts continue to be made in this regard. Theistic philosophers of religion, such as Norman Kretzman and Alvin Plantinga, are engaged in this project. Second, one can attempt to recover the *meaning* of eternity as expressed in metaphysical categories through postmetaphysical hermeneutics. This is the direction in which hermeneutical theologians such as Rudolph Bultmann, Paul Tillich, and Robert Scharlemann have worked.

Dividing these two possibilities is a reference to the phenomenology of lived, temporal experience as it appears in linguistic expression. Representatives of the first possibility tend not to include the reference to narrated experience with the same centrality of concern exhibited by representatives of the second possibility. In the first possibility, claims of eternity are established by logical inference of a transcendent eternal reality from more certain claims of knowledge. The exemplars of this method remain the classics of metaphysics from Plato and Aristotle to Anselm and Aquinas. They are united in the use of this metaphysical axiom: If one establishes a necessity of thought (in this case, the immortal soul participant in God's eternity), then one establishes therewith the existence of a necessary and universal reality. In the second possibility, claims of eternity are established by correlating phenomenological analysis of universal and necessary (hence "ontological") structures of human existence with interpretations of the religious meaning of eternity as expressed through linguistic symbols. Examples of this method of correlating Heidegger's existential analysis with interpretation of religious symbols include the post-Heideggerian theologians central to this project: Tillich, Bultmann, Ricoeur, and Scharlemann.

In the present circumstances of thought, there is no necessity to choose between these two possibilities. Despite other differences, both possibilities seek to articulate the human encounter with eternity. They have more in common with each other than either one has with competing modes of thought (such as empiricism or functionalism) that rule out in advance any encounter with eternity. However, this study pursues the second possibility while remaining open to the metaphysical option. Even if the metaphysical option were to become convincing today, the phenomenological-hermeneutical option still provides corroborating arguments with reference to eternity's appearance in language as opposed to thought. The key question is this: How can we understand God in God's eternality when our sense of eternity is no longer linked inexorably to an unquestionable metaphysical worldview? In other words, given that the status of metaphysics is unsettled today, can we retrieve the *meaning* of eternity expressed in traditional texts for confirmation with reference to interpreted structures of human existence?

To deny death its ontological status would seem to be equivalent to denying the force critical philosophy has played in calling into question metaphysical realities other than those necessary transcendental conditions of human knowing. We cannot simply rest content with a precritical metaphysics, wherein eternity reigned over death on the ontological horizon. The challenge to the traditional metaphysical perspective has not necessarily led to its demise, or to the demise of our religious heritage; it has, however, led to radical changes in the way in which we conceive theological reflection. Schleiermacher founded "modern" theology as a response to the new situation engendered by the Enlightenment and especially by Kant's critical philosophy. There have been significant theological responses to Heidegger as well. This study reviews some of the more important responses to Heidegger in chapter 3.

The desire to offer a theological response to Heidegger's interpretation of the ontological primacy of being toward death is not unique. Bultmann and Tillich have both attempted to respond theologically to Heidegger and to overcome existentially the primacy of death by reference to a postcritical interpretation of the Word of God. For Bultmann, it is the preached Word, the kerygma, that overcomes the sting of ontological death. For Tillich, it is the religious symbol as Word of God that does so. This study's proposal participates substantially in the existential theological responses of Bultmann and Tillich, but it is not identical with them. Let me indicate the elements of the particular problem with which I work, for it is not one directly taken up by either Bultmann or Tillich.

First, Bultmann and Tillich both had a primary interest in interpreting the meaning and power of New Testament symbols. In contrast to them, I am taking up the possibility that a classic theological text—Augustine's *Confessions*—can function in a

religiously symbolic way, even when purporting to offer conceptual treatment of time and eternity within the form of autobiography.

Second, Bultmann and Tillich both accepted the results of
Heidegger's analysis of Dasein when he concluded that temporality is the meaning of the being of Dasein. Each tried to show how
some supratemporal power of meaning could, however, break into
human temporality to create a form of faith or ultimate concern
grounded existentially not in temporal experience but in the
eternal horizon of temporal experience. For each of them, the
symbol marking the in-breaking of God is a traditional religious
image, whose primary referent is an object taken as a symbol of
the divine: namely, the cross of Christ. In contrast to Bultmann
and Tillich, I focus upon a temporal experience as a symbol of the
divine. Under consideration is the possibility that Augustine's
experience of "redemptive time" is the experience of a religious
symbol that has the power to convey a meaning that surpasses
ontological death as Heidegger understands it. The focus is not
upon the symbolic power of some image of an object in time but,
rather, on the symbolic power of a narrated account (in Augustine's *Confessions*) of a redemptive temporal experience.

There is no absolute difference between my analysis and those
of Bultmann and Tillich. To experience the cross as religious
symbol entails the experience of redemptive time. In phenomenological terms, the *noema* of the cross as symbol implies the *noesis*
of redemptive experience in time. But by virtue of focusing on
Augustine's narrated account of redemptive time rather than on
the New Testament symbol of the cross, the emphasis falls on this
noetic side. Because we read of Augustine's redemptive experience in his autobiography, his narrative account of redemptive
time can itself serve as a religious symbol for the readers of the
Confessions. The thesis being argued is that experienced redemptive time, as narrated in the *Confessions,* can function as a
religious symbol in Tillich's sense (even though Tillich himself
never entertained the notion of time as symbol). A theological
analysis of this symbol shows that it can overcome Heidegger's
understanding of the ontological primacy of death by revealing a
meaning that frees the mind from its fixation on death.

In carrying out this postcritical theological interpretation of
redemptive time in Augustine's *Confessions,* we are led to a

detailed study of Paul Ricoeur's three-volume *Time and Narrative* for three reasons. First, in this work, Ricoeur directly takes up the problem of time and eternity. He analyzes how Augustine constructs a solution to the question of time in relation to eternity and how Heidegger deconstructs it. In the contemporary literature, Ricoeur's interpretations of the crucial texts simply cannot be ignored.

> For Augustine and the whole Christian tradition, the internalizing of the purely extensive relations of time refers to an eternity where everything is present at the same time. . . . Yet Heidegger's philosophy of time, at least during the period of *Being and Time*, . . . orients its meditation not toward divine eternity but toward finitude sealed by being-towards-death [*Sein zum Tode*]. . . . The most serious question this work may be able to pose is to what degree a philosophical reflection on narrativity and time may aid us in thinking about eternity and death at the same time.[21]

Ricoeur's solution to the problem of how to speak of Augustine and eternity in light of Heidegger and ontological death stems from his criticisms of *Being and Time*. He places Heidegger's analysis of time (as well as Kant's and Husserl's) within the tradition of Western reflections on time that hearkens back to Aristotle and Augustine. His claim against Heidegger is that he does not pay sufficient attention to the cosmological strand of reflections on time that begins with Aristotle's *Physics*. That oversight leads to Heidegger's assumption that death alone is the limiting horizon for our understanding of time. By criticizing Heidegger's attempt to derive a cosmic sense of time from a phenomenological or subjective sense, Ricoeur makes room for a "fictive" version of Augustine's eternity. He argues that death and eternity can be brought together as the twin limiting horizons within a "narrative consciousness" that has historical and fictional components. Death limits the historical component of narrative consciousness, eternity the fictional component. But what happens to the religious sense of Augustine's eternity in Ricoeur's fictional interpretation? Is the living God whose Eternal Word spoke to (and through) Augustine's soul a mere figment of imagination? Ricoeur does not address the issue of being addressed *by* a voice in the text, but that is the intention of this study.

Second, Ricoeur makes explicit a problem facing anyone who reflects on time, although it has largely gone unreflected hitherto. The problem is that in reflecting on time a peculiar dilemma poses itself: Time divides into "psychological time" and "cosmological time." Prior accounts of time focus on either one or the other but do not attempt to analyze and explain the relation between the two forms time can take for reflection: lived time and measured time. Ricoeur's work breaks this pattern. He analyzes the texts in which the dilemma appears and proposes a solution to it. According to Ricoeur, the only time upon which we can reflect is "narrated time." Reflected time is narrated time, whether described psychologically or cosmologically. Ricoeur claims that the notion of narrated time is, in fact, the hidden ground of the difference between psychological and cosmological time: It is that point in reflection from which we can see both the identity and the difference of these forms of time. In order to reflect on "experienced redemptive time" as it appears in Augustine's *Confessions,* I shall reflect on it as "narrated time," following Ricoeur's analyses. This necessitates my analytical exposition of *Time and Narrative.*

Third, in making a theological interpretation of experienced redemptive time, this study moves beyond what Ricoeur himself analyzed in *Time and Narrative.* However, in order to theologically interpret experienced redemptive time as narrated, we must first examine the difficult arguments in support of the notion of narrated time in *Time and Narrative.* Particularly interesting are Ricoeur's own few suggestions about the meaning of eternity construed nonmetaphysically through the narrated time of the Hebrew Bible. Ricoeur speaks about "fidelity" as such a nonmetaphysical meaning of eternity in his "Conclusion." In making a theological interpretation of narrated time, I shall apply Robert Scharlemann's suggestive comments in the essays herein cited.

Scharlemann argues that Ricoeur's hermeneutical theory lacks an appreciation for how religious texts can manifest a world in which death has been relegated to the past. In certain religious texts, the future unfolds an eternal horizon that is beyond death. Scharlemann claims that Ricoeur's hermeneutics falls short religiously because it lacks a notion of "textuality." Textuality refers to the unique identity and voice of a text, which enables it to present a world for possible habitation. Due to the linguistic powers of a text's textuality, the world it presents need not always be a redescription of our existential-historical world. But according to Scharlemann, Ricoeur's hermeneutics only allows for redescriptive interpretation because it lacks a notion of textuality. By providing for that lack, Scharlemann is able to extend Ricoeur's hermeneutics religiously and theologically. I employ Scharlemann's argument by indicating how the textuality of Augustine's *Confessions*—that is, the eternal voice of God that speaks through Augustine from beyond death—can actually be discovered in certain religious hints that Ricoeur makes in the "Conclusion" to *Time and Narrative.* To the extent that we can still hear the eternal as a voice of textuality in Augustine's text and understand what it means, we need not give up the full religious sense of eternity as an aspect of God's being. Turning to Scharlemann helps remedy the lack of theological perspective in Ricoeur's confrontation between Heidegger and Augustine.

Chapter Outlines

The argument of this study is divided into a two-step strategy. The first step is to follow Ricoeur's argument in *Time and Narrative* up to the point where it begins to open itself to theological evaluation and extension. This is accomplished in chapters 1 through 3 herein. The second step is to provide a religiously significant response to ontological death by applying Scharlemann's religious extension of Ricoeur's hermeneutics to the latter's reading of the *Confessions.* This is accomplished in chapter 4. The second step fulfills the inquiry into time, death, and eternity that is initiated through an extended, critical conversation with Ricoeur. Let me be more specific about how these steps unfold in the chapters that follow.

Chapter 1, "Saint Augustine's *Confessions,*" accomplishes two things: (1) it initiates the first step of the argument by reviewing Part I of Ricoeur's *Time and Narrative;* and (2) it begins laying the groundwork for the thesis that the *Confessions* can still speak to us as the eternal voice of God despite the threat to religious meaning engendered by the increased ontological significance of death. In Part I of *Time and Narrative,* Ricoeur compares Augustine's *Confessions* to Aristotle's *Poetics.* Through a conjoint reading of these classic texts, he develops his thesis: In order to resolve the aporia that arises whenever we think about the nature of time (i.e., the inability of the mind to grasp all the features of time in one conceptual construct), we must rely upon the poetic powers of imagination that can transform the disjunctions of our temporal experience into a coherent, narrative scheme. Ricoeur assumes that the problems that Augustine runs into in his reflections on time (Book XI) are paradigmatic for any thinking about such a phenomenon. We cannot say *what* time is philosophically because we cannot resolve the discrepancies in point of view (the aporias) that constantly crop up whenever we think of time. That is the aporetical problem: Time entails aspects that embrace both sides of a series of logical contradictions. For example, Augustine is forced to argue that time is both existent and nonexistent; that it is both measured and the measure of things; and that it is both unifying (intentional) and distending.[22] According to Ricoeur, the harder Augustine works to figure out time, the more problems he creates for himself because of time's aporetical nature. The only way to resolve this problem is to allow the productive imagination to construct a narrative that can make productive use of the contradictions inherent to temporal experience.

Ricoeur develops Aristotle's concept of "mimesis" in offering a poetic response to philosophical quandaries about time. Mimesis is the imitation of action or life experience that the imagination accomplishes in constructing a narrative plot. The mimetic process extends to the level of text the same abilities to create new meaning that the metaphoric process was seen to accomplish on the sentence level in Ricoeur's earlier work, *The Rule of Metaphor*.[23] According to Ricoeur's tension theory, metaphors overcome contradictions of the literal sense of sentences by creating new possibilities for meaning upon the nonsense of the contradictions. In a similar way, mimetic plots can overcome the contradictions implied in the aporias of temporality by interweaving their tensions into a coherent narrative.

After reviewing the argument in Part I of *Time and Narrative*, I offer an original reading of Book XI of the *Confessions*. This helps lay the groundwork for the theological extension of Ricoeur's last reading of the *Confessions* (given in his "Conclusion") in the second step of the overall strategy to respond to ontological death. My reading of the *Confessions* is critical of Ricoeur's reading on one basic point: I argue that Ricoeur tends to overlook the religious context of Augustine's discussion of time and eternity. The context is an interpretation of Genesis 1:1. Augustine wants to know what the difference is between Creator and creature. The eternity/time dialectic is developed in order to explain that difference. Rather than call Augustine's treatise on time a psychological theory, as Ricoeur does, I prefer to call it a "redemptive theory." This helps preserve the obvious religious importance of Augustine's argument. Criticizing Ricoeur on this score anticipates the extension of his interpretation that is accomplished in chapter 4.

Chapter 2, "From Aristotle and Augustine to Heidegger: Reflecting on Time in the West," continues to fulfill the first step in my two-step strategy to respond to ontological death. Chapter 1 focused on Part I of *Time and Narrative,* but now the focus shifts to Part IV. In Part IV, Ricoeur fulfills the promise that he made earlier: to spell out a narrative response to philosophical quandaries about time. First he reviews the history of philosophical reflections on time in the West, focusing on Aristotle's *Physics,* Augustine's *Confessions* (once again), Kant's *Critique of Pure Reason,* Husserl's *Phenomenology of Internal Time-Consciousness,* and Heidegger's *Being and Time.* Ricoeur argues that the same aporetical quandaries that plagued Augustine's attempt to think the nature of time plagued all the other classics in this tradition as well. He discovers that there is one fundamental aporia that is at the root of all the others. It is the unresolvable discrepancy between a phenomenological sense of time and a cosmological sense of time. In the former case, time is viewed from the perspective of the human subject; in the latter case, from the perspective of nature or the cosmos. Ricoeur demonstrates how this aporia divides the philosophical tradition into two camps, one stemming from Aristotle, the other from Augustine. Aristotle's *Physics* stands at the beginning of a cosmological tradition of reflections on time, which includes Kant's *Critique* and also contemporary scientific theories about time. Augustine's *Confessions,* on the other hand, stands at the beginning of a phenomenological tradition of reflections on time, which includes Husserl's *Phenomenology* and Heidegger's *Being and Time.* Ricoeur argues that a third sense of time—narrated time, the time of narrative (or historical) consciousness—is necessary in order to mediate the subject-object split that indwells our thinking about time.

After interpreting the tradition of reflections on time in terms of the phenomenology/cosmology aporia, Ricoeur demonstrates that narrative understanding already entails the interweaving of phenomenological and cosmological components. He argues that every narrative has fictional and historical aspects, which correspond to the phenomenological and cosmological strands of philosophical reflection. By interweaving the fictional and the historical, narrative consciousness is able to achieve a new sense of time—narrated time—that both differs from phenomenological time and cosmic time and overcomes their differences. Thus, Ricoeur's argument for narrative consciousness, and its sense of time that mediates the basic philosophical aporia, fulfills his project to think time and narrative together. But what of his other claim, that the most serious question his text raises is how to think death and eternity together? My concern in chapter 2 is to begin articulating precisely how Ricoeur mediates the apparent discrepancy between death and eternity (or between Heidegger and Augustine).

The problem of death and eternity is embedded within the problem of narrative consciousness—namely, how to overcome the philosophical aporia that splits our thinking about time. Even though Ricoeur places Augustine and Heidegger within the same phenomenological strand of reflections on time, he distinguishes them by determining which one stands closer to the cosmological strand. Heidegger's claim that death is the determining factor (or hierarchizing principle) for human temporality places him closer to the cosmological sense of time, according to Ricoeur. On the other hand, Ricoeur thinks Augustine's claim that eternity is the determining factor for human temporality places him closer to a true phenomenological sense of time. Ricoeur argues that Heidegger makes a mistake when he takes a historical fact (we all must die) for a subjective or phenomenological principle (my being is determined solely by *Sein zum Tode*). To take the historical (which is closer to a cosmological sensibility than to a phenomenological one) as determinative for all thinking is too simplistic. It precludes the possibility that our thinking can legitimately be extended—through imagination—beyond the strict limits of historical experience. Kant allowed for that possibility in the *Critique of Pure Reason* and other works, and Ricoeur insists that Heidegger should not have overlooked such a possibility. A true phenomenological perspective, according to Ricoeur, is linked to the imagination and its abilities to extend thinking beyond strict historical limits. Thus, he interprets Augustine's eternity as a Kantian limit-idea (or religious symbol) that imaginatively extends our thinking beyond Heidegger's *Sein zum Tode.* Within narrative consciousness, which mediates the two poles of our thinking, eternity and death are brought together. Eternity stands as the limiting horizon for the fictional component of narrative consciousness, which is able to extend phenomenological thinking beyond historical limits. Death stands as the limiting horizon for the historical component of narrative consciousness, which serves to confine our thinking to cosmological constraints.

So the first section of chapter 2 focuses upon Ricoeur's project to think death and eternity together. But then I turn back to Ricoeur's treatment of the philosophical tradition of reflections on time in order to follow his moves from Aristotle and Augustine up to Heidegger more carefully. This helps establish just how Ricoeur relates the *Confessions* to *Being and Time.*

Chapter 3, "Heidegger's *Being and Time,*" continues to fulfill the first step of my argument. I offer an original reading of Heidegger's text as I interact with Ricoeur's interpretation. In order to substantiate the claim that *Being and Time* fulfills a tendency in Kant's critical philosophy, Heidegger's argument in *Kant and the Problem of Metaphysics* is reviewed. Ever since Kant's *Critique of Pure Reason,* the status of time has increased as the status of a metaphysical, eternal realm has decreased. With the publication of *Being and Time,* we see this tendency fulfilled: When death—time's principal consequence in human life— becomes the determining factor for our being in the world and self-understanding, then the tendency for criticism to displace transcendence has reached a terminal point. Death is the opposite of transcendence; its ontological ascendancy corresponds to the decline of transcendent possibilities. Religious meaning is threatened by the ontological death depicted in *Being and Time.* In *Kant and the Problem of Metaphysics,* Heidegger demonstrates how his philosophy of finitude grows out of an interpretation of the *Critique of Pure Reason.*

At the end of chapter 3, I offer a brief review of the impact

Heidegger has had on theology. This helps shift the argument toward its theological outcome. Chapters 1, 2, and 3 are devoted primarily to the task of following Ricoeur's efforts to think death and eternity together. Once that is accomplished, the next step is to extend Ricoeur's argument theologically so that the religious significance of Augustine's eternity can be preserved, as well as its philosophical significance. The move in that theological direction begins with a discussion of Heidegger's influence on specific theologians, particularly, Bultmann, Tillich, and Scharlemann. I end the chapter with a discussion of Scharlemann's "The Being of God When God Is Not Being God." In that essay, Scharlemann indicates how theologians, following the example of Heidegger, can accomplish a deconstruction of the metaphysically driven "theistic picture" that continues to dominate Western thought. He claims that the kind of deconstructive rereading of the philosophical tradition that Heidegger accomplished in *Being and Time* has yet to be duplicated for the theological tradition. He argues that the otherness of God has remained unthought and conceptually forgotten in theology, just as, according to Heidegger, the question of the meaning of being has remained unthought and conceptually forgotten in philosophy. Turning to Scharlemann's essay on Heidegger and theology anticipates the second step of the strategy to respond to Heidegger's understanding of ontological death. What remains unthought in Augustine's text is how redemptive time is the being of God (as eternal) when God is not being God. As I shall indicate in a moment, the final reading of the *Confessions* that Ricoeur offers (in his "Conclusion") is deconstructive in Scharlemann's (Heidegger's) sense. Ricoeur attempts to isolate the Hebraic qualities of Augustine's reflections on time and eternity from their Greek metaphysical trappings. Even though this deconstruction is never carried through, it marks a theological turn in Ricoeur's thought. I employ Scharlemann's formula for deconstruction in order to fulfill the reading of the *Confessions* that Ricoeur only begins. This fulfillment takes place in chapter 4, but it is anticipated in the review of "The Being of God When God Is Not Being God" at the end of chapter 3.

Chapter 4, "*Time and Narrative*: A Theological Extension," accomplishes the second step of the strategy. It begins with a review of Ricoeur's "Conclusion" to *Time and Narrative,* where he initiates the deconstructive reading of the *Confessions* I mentioned above. Ricoeur claims that the Hebrew sense of eternity—God's eternal fidelity—is the mythic archaism that shapes and informs Augustine's discussion of eternity as the other of time. I employ Scharlemann's strategy for deconstruction in order to bring Ricoeur's final reading of the *Confessions* to its proper conclusion. Before that, however, I introduce another essay by Scharlemann, the principal one in this theological extension of Ricoeur's reading of the *Confessions:* "The Textuality of Texts." In this essay, Scharlemann criticizes *Time and Narrative,* and Ricoeur's hermeneutics in general, for its inability to appreciate religious texts that present a world beyond the care and death of our historical existence.

Employing both essays by Scharlemann—one on deconstruction, the other on Ricoeur's hermeneutics—enables me to extend Ricoeur's reading of the *Confessions* and to claim that a religious understanding of this text, particularly the discussion of eternity and time, is still possible. The textuality of the *Confessions,* its unique voice, bespeaks a world where death is already in the past. It is the voice of God in God's otherness calling us home. Ontological death, therefore, has not eclipsed religious meaning.

Notes

1. See Karl Jaspers's *The Atom Bomb and the Future of Man* for a convincing discussion of the "new fact" that the bomb forces upon the human race.

   Edith Wyschogrod, in *Spirit in Ashes: Hegel, Heidegger, and Man-Made Mass Death,* argues that experiences such as Auschwitz and Hiroshima have placed us in a new death situation. The situation of "man-made mass death" has rendered the old pattern for facing death—the "authenticity paradigm"—obsolete. Within the authenticity paradigm, the meaning of a person's life, and its moral worth, was tied to how well or gracefully that person faced death. Yet the concentration camp and technological warfare—both all too common in this century—deprive us of the familiar life-world that is necessary for the authenticity paradigm to work properly. We face a new kind of threat in the "death-world" of man-made mass death. Wyschogrod argues that the death-world is not identical with technological society, though it is nevertheless an unfortunate consequence of it. Following the later Heidegger, she depicts the dehumanizing forces inherent within technology—forces that can lead to the death of our caring existence, our life-world.

2. One of the important bioethical questions, particularly for the medical and legal professions, concerns the proper medical criteria for determining death. Traditional criteria for pronouncement of death have been absence of pulse and of respiration. "Recently, however, technological advances have made it possible to sustain brain function in the absence of spontaneous respiratory and cardiac function, so that the death of a person can no longer be equated with the loss of these latter two natural vital functions. Furthermore, it is now possible that a person's brain may be completely destroyed even though his circulation and respiration are being artificially maintained by mechanical devices" (Frank Veath et al., "Brain Death," in *Bioethics,* edited by T. Shannon, 3d ed. [Mahwah, NJ: Paulist Press, 1987] 171–72). After reviewing the issues from clinical, philosophical, religious, legislative, and judicial perspectives, Veath, et al. recommend a particular definition of brain death as a new criterion for determining death.

The point is that new technical skills are forcing us to rethink completely our understanding of the phenomenon of death. And these new skills have led some prophets of "cryonic hope" to view death as an unacceptable imposition on the human race! (See chapter 4, "The Changing Sociology of Death," in John Hicks's *Death and Eternal Life,* for a thoughtful discussion of "cryonic hope" and other aspects of our new predicament).

3. Elisabeth Kübler-Ross, in her famous *On Death and Dying,* suggests that we have more fear of death than did previous generations in part because we are forced to die in unfamiliar surroundings—in the hospital instead of in the home. Her mission, as psychiatrist, has been to allay some of these increased anxieties.

4. See chapter 3 herein for an extensive discussion of *Being and Time,* particularly the notion of *Sein zum Tode.*

5. *Critique of Pure Reason,* A334.

6. See Scharlemann's discussion of the senses of "God Is" in *The Being of God,* 47–93.

7. See Gadamer's *Truth and Method,* 91–118.

8. See volume 3 of Hegel's *Lectures on the Philosophy of Religion,* 173–89, 351–58.

9. *Dynamics of Faith,* 41–54.

10. See Tillich's article "The Word of God," in *Language: An Enquiry into Its Meaning and Function.*

11. In *The Courage to Be,* Tillich claims that the ancient world was preoccupied with death and thus was plagued with "ontic" anxiety. He claims further that the modern era is plagued with spiritual anxiety due to the loss (or threatened loss) of the ultimate or transcendent as such. It might seem that my argument for a concept of ontological death contradicts Tillich's scheme, but I do not think so. The solution to ontic anxiety in the ancient world was given by Socrates and Jesus, both of whom advocated hope in the transcendent world to come after death. Today, however, we do not have such confidence in the possibilities for eternal life. Spiritual anxiety reflects this loss of confidence. What coincides with a loss of the transcendent (Tillich's concern) is a rise in the ontological significance of death (my concern). The ontic anxiety of the ancient world has become, for us, an "ontological" (spiritual) anxiety because the ultimate significance of death has increased as belief in the possibility of afterlife has decreased.

12. In "The Textuality of Texts," Scharlemann notes how commitment to an axiom of the finality of death was one significant consequence of the rise of a critical historical consciousness. Such an axiom is evidence for the rising ontological significance of death. For a

further discussion of Scharlemann's article, see "Beginning a Theological Extension" in chapter 4 herein.

13. I realize that I am oversimplifying the history of philosophy by suggesting a direct link between Kant and Heidegger. In "From Kant to Heidegger" in chapter 3, I give a more responsible account for the move from Kant to Heidegger that I am asserting here.

14. See "Western Classics on Time," in chapter 2 herein for a more extensive discussion of Kant's *Critique of Pure Reason*. Here, I am only trying to give a general orientation to the relevant issues.

15. See Plato's "Phaedo" for a discussion of his theory of recollection in *Plato: The Collected Dialogues,* 40–98.

16. The same idea is present in Ricoeur's hermeneutical theory, which develops Heidegger's analysis of understanding and interpretation: Whenever we understand something, it is through the development of an originary preunderstanding. Without a preunderstanding of our being in the world, we would not be able to recognize or "know" anything. Heidegger's notion of world also serves a function similar to that of Kant's transcendental time determinations (especially since they are the basis for an originary unity of apperception) and that of Plato's forms. In *The Hermeneutical Theory of Paul Ricoeur,* David Klemm argues that what Heidegger and Ricoeur call *Verstehen,* or understanding, takes the place of Kant's transcendental imagination. The transcendental time determinations are a product of the transcendental imagination.

17. I realize that I am overlooking Heraclitus and other philosophers of becoming. But theirs was never the dominant view in Western thought.

18. See 1 Cor. 15:51ff., where Paul, speaking of the Parousia to come, says: "Lo! I tell you a mystery. We shall not all sleep, but we shall all be changed, in a moment, in the twinkling of an eye, at the last trumpet. For the trumpet will sound, and the dead will be raised imperishable, and we shall be changed. For this perishable nature must put on the imperishable, and this mortal nature must put on immortality. When the perishable puts on the imperishable, and the mortal puts on immortality, then shall come to pass the saying that is written: 'Death is swallowed up in victory.' 'O death, where is thy victory?' 'O death, where is thy sting?' "

19. There is a tradition of secular acceptance of death that would not feel a loss in the idea of *Sein zum Tode*. In the modern period, this tradition extends from Hume through Russell and also includes many psychological thinkers; among the ancients, some of the Stoics and Epicureans fostered this attitude. But it seems to me that many of us, Heidegger included, feel the loss of transcendence very

deeply. I shall be concerned with only those who do take this loss to heart.

Furthermore, I think the sense of sadness and loss many of us feel because of death's ascendancy to a higher ontological plateau affects our attitude toward the other new circumstances we face today. Ontological death probably redoubles the existential anxiety and despair we may feel concerning nuclear weaponry and medical technologies. If, for instance, our religious hope is already threatened by ontological death, then it becomes easier for us to feel skeptical and complacent about nuclear arms. Lacking any significant hopes for this life, we can more easily say, "So what if we drop the bomb and the whole world goes up in smoke?" Why care about life if death is the only outcome? Ontological death can lead to the sort of desperation that increases the likelihood of a nuclear holocaust.

Hannah Arendt, in *The Origins of Totalitarianism,* argues a somewhat similar point with respect to the extermination camps of Nazi Germany. She claims that loss of religious faith in a final judgment made the concentration camps—here-and-now expressions of an everlasting hell—possible. The first step on the road to totalitarian ideologies is the loss of the inner, juridical self for whom final judgment and hell are imaginable only in light of the infinite possibility of grace. Radical evil is an unprecedented concept in Western tradition, and yet it has become a horrible reality. According to Arendt, "It is inherent in our entire philosophical tradition that we cannot conceive of a 'radical evil,' and this is true both for Christian theology, which conceded even to the devil himself a celestial origin, as well as for Kant, the only philosopher who, in the word he coined for it, at least must have suspected the existence of this evil even though he immediately rationalized it in the concept of a 'perverted ill will' that could be explained by comprehensible motives. Therefore, we actually have nothing to fall back on in order to understand a phenomenon that nevertheless confronts us with its overpowering reality and breaks down all standards we know. There is only one thing that seems to be discernible: we may say that radical evil has emerged in connection with a system in which all men have become equally superfluous" (459). Loss of hope in a transcendent afterlife, which, in its Christian expression, includes a sense of individual judgment before God, has left us vulnerable to unprecedented manifestations of human behavior.

With respect to medical technologies and bioethics, one might ask, Why is there such hesitancy to "pull the plug" on patients such as Karen Ann Quinlan whose lives are being artificially sustained?

What are we trying to attain through medical science? It might be that the loss of a transcendent realm has put key pressure on health care. Perhaps doctors are required to compensate for this loss by assuring us the longest possible life span. To pull the plug is like giving up belief in eternal life; subconsciously, we may view medicine as our last line of hope against ontological death. Such a view is supported by claims that Arendt makes in *The Human Condition*. She says that life as such became the highest good in Western societies when Christian belief in the immortality of the individual soul became the central creed (see 316ff.). But now that Christian belief in the world to come is no longer so viable, what we are left with is an empty conviction that life must go on at all costs. Hence our hesitancy to "pull the plug."

And why do we tend to deny death more than past generations have? This is a question that Kübler-Ross ponders, as does Ernest Becker in *The Denial of Death*. I would suggest that we deny death more because the stakes are higher. That death has achieved greater ontological significance, at the general expense of religious hope in a world to come, has made its presence more oppressive to bear. Like the hospital doctors, we all want to fight against such a ruthless foe. The problem is that when we deny death too much, it leads to mental illness, which is Becker's point (and Kübler-Ross's). For us, Pascal's fear of an indifferent universe has come home to roost in the realization that life is nothing but a *Sein zum Tode*. Somehow we must learn to live with the increased anxiety.

20. Early in the nineteenth century, the romantics asked this question and were the first to begin recovering a nonmetaphysical sense of the "eternal now." Compare, for example, Wordsworth's "spots of time" (Book XI, 257–78, 1805 Prelude, *The Prelude: 1799, 1805, 1850*) and Schleiermacher's interpretation of eternity in the *Soliloquies*.

21. Ricoeur, *Time and Narrative* (I, 86–87). The roman numeral preceding page numbers for Ricoeur's quotations from *Time and Narrative* refer to the volume number of this three-volume work, not the part number, throughout this text.

22. In Part IV of *Time and Narrative,* Ricoeur argues that the most important aporia with respect to time has to do with the fact that it is a phenomenon of both the psyche and the cosmos. This is the basic discrepancy within our thinking the nature of time that gives rise to a host of aporetical quandaries whenever we try to think time as such.

23. See chapter 4 herein for a more detailed discussion of how the mimetic process is an application of the metaphoric process, which is itself a poetic application of Kantian schematism.

Chapter 1
Saint Augustine's Confessions

> *Let us come home at last to you, O Lord, for fear that we be*
> *lost. For in you our good abides and it has no blemish, since*
> *it is yourself. Nor do we fear that there is no home to which*
> *we can return. We fell from it; but our home is your eternity*
> *and it does not fall because we are away.*
>
> *Confessions,* Book IV, ch. 16, p. 90

Ricoeur offers three different readings of Augustine's autobiography in the progression of *Time and Narrative.* The difference between the first two readings and the third is the most significant. His first and most thorough reading is offered in Part I, which is the section I focus on in this chapter. In Part I, Ricoeur creates the context for his entire argument about time and narrative by starkly contrasting Augustine's "aporetical" interpretation of time (Book XI, *Confessions*) and Aristotle's mimetic interpretation of a narrative plot (*Poetics*).[1] Later, in Part IV (the focus of my next two chapters), Ricoeur's contrast between Augustine and Aristotle shifts ground. Augustine's psychological sense of time is contrasted with Aristotle's cosmological sense of time (*Physics*). Ricoeur's purpose in Part IV is to give a "narrative" account of the significant discussions of time in Western philosophy. Augustine and Aristotle provide the basic *agon,* or tension, that is repeated in Kant, Husserl, and Heidegger and that must be overcome narratively. Later still, in his "Conclusion" (which I discuss in detail in chapter 4), Ricoeur alludes to the theological significance of Augustine's discussion of time in its relation to eternity or the "other" of time. Here, the contrast between Augustine and Aristotle shifts dramatically as Ricoeur attempts to distinguish a biblical sense of eternity (Augustine) from a Greek sense (Aristotle, Plato).

31

This chapter is divided into two parts. First, a brief overview of Ricoeur's argument in Part I of *Time and Narrative* is given. Second, an original reading of Book XI of the *Confessions* is offered, comparing it step by step with Ricoeur's reading. I criticize Ricoeur for minimizing the theological context of Augustine's discussion of time. To a great extent, he overlooks the biblically inspired creation/redemption scheme that determines Augustine's eternity/time dialectic. Thus he calls Augustine's theory of time a "psychological" theory. But it would be more proper to call it a "redemptive" theory of time because it concerns the time when God enters the soul as a redemptive, eternal voice.

Time and Narrative, Part I:
The Circle of Narrative and Temporality

The principal thesis of *Time and Narrative* is set forth in Part I. Ricoeur argues that "time becomes human time to the extent that it is organized after the manner of a narrative; narrative, in turn, is meaningful to the extent that it portrays the features of temporal experience" (I, 3). He works his way toward this thesis of the reciprocity between narrativity and temporality in two stages. First, he presents Augustine's theory of time (Book XI, *Confessions*) and Aristotle's theory of emplotment (*Poetics*) as independent analyses that seem in themselves to give no indication of the thesis of reciprocity. Augustine says nothing about narrative in Book XI of the *Confessions,* and Aristotle says nothing about time in the *Poetics.* But in a second stage, Ricoeur shows how the confrontation between the *Confessions* and the *Poetics* can lead to his thesis after all. As Ricoeur reads them, the two texts represent inverted images of each other.

> The Augustinian analysis gives a representation of time in which discordance never ceases to belie the desire for that concordance that forms the very essence of the *animus.* The Aristotelian analysis, on the other hand, establishes the dominance of concordance over discordance in the configuration of the plot. It is this inverse relationship between concordance and discordance that seemed to me to constitute the major interest of a confrontation between the *Confessions* and the *Poetics.* (I, 4)

What does Ricoeur mean by claiming that the *Confessions* and the *Poetics* are inverted images of each other? The key contrast is between Augustine's torn or distended soul and Aristotle's mimetic emplotment. What the former lacks—namely, a sense of synthetic unity or concordance—the latter can provide. When the two themes of time and narrative are brought together, new insights into both are engendered. In what follows, then, I shall focus on the Augustinian *intentio* that causes *distentio* and the Aristotelian mimesis that poetically synthesizes the temporal components of the Augustinian opposition.

"*Intentio*" refers to the attention of the soul, which is always temporally determined; that is, there is attention to the past (in memory), attention to the future (in expectation), and attention to the present. The incommensurability among these three different attentions—one cannot attend to the past and the future at the same time—results in a distension of the soul. The soul is torn back and forth between the temporal horizons of its own intentions. The more we intend something, the more we are distended due to our temporal nature. This aporetical or antinomous situation is not easily resolved. The only way to resolve this state of distension, according to Ricoeur, is to construct a narrative plot that can interweave temporal differences. Mimesis constitutes the process of overcoming the discordance of *distentio* through the construction of a plot. Mimesis is actually an extension of what Ricoeur calls the "metaphoric process" (*The Rule of Metaphor*), which is itself a poetic extension of Kantian schematism. The productive imagination synthesizes the discordant components of our temporal experience through the process of mimesis. Thus, the soul's distension in Augustine's understanding and its mimetic synthesis in Aristotle's understanding are inverted images of each other. Let me explain in more detail.

Ricoeur begins his reading of Book XI of the *Confessions* at the point where Augustine asks, "What, then, is time?" (ch. 14.17, p. 263) Initially, Ricoeur isolates the discussion of time from its backdrop in a meditation on eternity. This is done in order to highlight the aporetical character of any pure reflection on time. An aporia is a disjunction in thought that cannot be resolved. As in a Kantian antinomy, two equally plausible possibilities become apparent. In Augustine's case, the discrepancy between *intentio* and *distentio* constitutes one of many aporias. The more the soul intends something, the more it suffers distension. The two perspectives cannot be resolved theoretically. We have to include both in order to understand the whole phenomenon of the temporal soul. Yet the aporetical nature of the discussion eventually has a negative effect: The ultimate outcome of an aporetical discourse such as Augustine's is a loss of concordance or sense of unified understanding. Ricoeur's point is that the more Augustine tries to isolate time as such, the more difficulties he uncovers. We have no sense of time in itself that can escape aporetical rhetoric. This impasse in the aporetical for any theoretical reflection on time, best illustrated in the *Confessions* and still valid in Husserl's and Heidegger's phenomenologies, enables Ricoeur to argue for his reciprocity thesis: "Speculation on time is an inconclusive rumination to which narrative activity alone can respond" (I, 6). There can be no theoretical resolution to the aporias of time (i.e., we can never put a precise finger on what time is and is not), only a poetical one.

Augustine's *Confessions,* as Ricoeur reads them, point to our discordant experience of time. Time is something that escapes our grasp because it marks the very flow of life. Neither life nor time can be easily stopped and bottled into the understandable confines of an idea. They flow on and on, marking changes, surprises, the unconfinable. Yet in Aristotle's concept of emplotment (*muthos*), Ricoeur finds a triumph of concordance over discordance. Narrative emplotment entails a supple logic that can unify the reversals and discontinuities found in life and time. (In chapter 4, I provide a full explanation of Ricoeur's poetic extension of Kant's schematism in his discussions of metaphor and mimesis.) The "logic" of mimesis is ultimately derived from Kant's imaginative variations of time in the synthesis of intuitions and concepts. In the case of mimesis, however, the synthesis is between lived temporal experience and human discourse rather than intuitions and concepts.

In reading Augustine and Aristotle together, Ricoeur establishes a relationship "between a lived experience where discordance rends concordance and an eminently verbal experience where concordance mends discordance" (I, 31). *Time and Narrative* is an inquiry into the mediating (schematizing) operations between lived experience and discourse. Through creative imitation of lived temporal experience (mimesis), a plot can render discordances concordant (i.e., it can overcome the dissipation of *distentio* as a result of *intentio*). Or, plots function as very flexible categories that somehow bring the indeterminate features of life and time under control. Life and time, in themselves aporetical and elusive to the point of mystery, are nevertheless subject to the rules of human interpretation through the special features of narrative discourse.

"The imitation of action is the plot" (50a1) is the text that guides Ricoeur's reading of the *Poetics*. From it, he extracts the two concepts he intends to borrow from Aristotle: mimesis and emplotment. Mimesis is a poetic activity—the imitation or representation of human actions—that produces a plot, which is the organization of events pertaining to imitated actions. Both mimesis and plot are poetic *operations,* not structures. Ricoeur emphasizes the productive, dynamic sense of Aristotle's understanding of "poetics." He considerably expands both concepts (mimesis and emplotment) beyond their Aristotelian senses.

Ricoeur articulates his mediation between time and narrative, or between Augustine and Aristotle, by further developing the mimetic structure that he extracts from the *Poetics*. The dynamic of emplotment, he says, is "the key to the problem of the relation between time and narrative" (I, 52). Ricoeur wants to show

> the mediating role of the time of emplotment between the temporal aspects prefigured in the practical field and the refiguration of our temporal experience by this constructed time. *We are following therefore the destiny of a prefigured time that becomes a refigured time through the mediation of a configured time.* (I, 54)

Thus Ricoeur isolates three distinct moments in the mimetic process.[2] Mimesis$_1$ is the transposition of real-life experience into the imaginary, poetic domain. It is comparable to the point in Kant's interpretation of the knowing process whereby a manifold of intuitions—a sense of our experience in the world—is made available to the imagination for synthesis with concepts. Mimesis$_2$ is the actual synthesis or imaginative invention of a new plot. The plot is conceived and constructed as the imagination attempts to imitate ethical actions in the world. These actions play the normative role of the schematized categories in Kantian synthesis. Mimesis$_3$ marks the completion of the poetic refiguration of the real by projecting a cultural world within which readers can be affected. The persuading, cathartic powers entailed in narrative configurations come to fruition in the spectator or reader. Thus, mimesis$_2$ (the configuring of a plot) plays a mediating role in the mimetic process. It connects preceding and succeeding stages of practical-temporal experience, which is indicative of the mediation between time and narrative.

Mimesis$_1$, the preunderstanding of the world of action that grounds the composition of the plot (mimesis$_2$), has three principal features. The first feature is a structural one. A practical understanding of the actions and suffering of agents contributes the structural building blocks (e.g., action sentences) with which narratives are constructed. Narrative understanding (mimesis$_2$) presupposes a practical understanding (mimesis$_1$), which it transforms and integrates into the order of the plot.

The second feature that anchors emplotment in our practical understanding has to do with the fact that actions are always already symbolically mediated. An implicit symbolism that articulates cultural rules and norms confers an initial readability upon actions that cannot be ethically neutral. This initial symbolic reading of actions, with its inherent ethical quality, is further developed in the poetic interpretation that engenders a plot.

The third feature of mimesis$_1$ is the most pertinent, in that it "concerns the temporal elements onto which narrative time grafts its configurations" (I, 59). Ricoeur indicates the temporal bridge between the practical and narrative realms (or between mimesis$_1$ and mimesis$_2$) with the help of Heidegger's notion of within-time-ness (*Innerzeitigkeit*).[3] Due to the within-time-ness of our care, or the fact that our being is within time, all our actions in the world can be characterized in terms of preoccupation. The temporal expression of our care is preoccupation: It is the time to do something, the "now" within which the three "ecstases" or distensions of time (past, present, future) are united in action. Narrative configurations, like the care structure of our being and acting in the world, share the same foundation of within-time-ness. This within-time-ness is the final bridge between mimesis$_1$ and mimesis$_2$.

Mimesis$_2$ is the configuring activity of emplotment. It mediates between a preunderstanding and a "postunderstanding" of the order of action and its temporal features. Ricoeur claims there are at least three modes of mediation. First, a plot mediates individual events and the story taken as a whole. "Emplotment is the operation that draws a configuration out of a simple succession" (I, 65). Second, emplotment brings together heterogeneous factors—agents, goals, means, interactions, circumstances, unexpected results, and so on—such that a concordance emerges out of discordance (or such that meaning emerges out of chaos). The third mode of mediation is related to the first and the second, and it involves the temporal characteristics of a plot. Two temporal dimensions are involved in the act of emplotment, one chronological and the other not chronological. The former constitutes the episodic dimension of the plot, the latter the configurational dimension. The episodic characterizes the story as a succession of events; the configurational characterizes it as a whole. The configurational act of emplotment grasps together the events or incidents of a story, forming a unified temporal whole, much as a synthetic judgment, according to Kant, grasps together an intuitive manifold under the rule of a concept. The act of emplotment, like a judgment, extracts a configuration from a succession. The two temporal poles of emplotment (episodic event and configured whole) reflect the Augustinian paradox of time (i.e., the incommensurability between distension and intention). Yet in the mediation of the two poles, the act of emplotment—extracting a figure from a succession—initiates a poetic solution to the paradox. This solution is revealed to the reader in the story's capacity to be followed.

> It is this "followability" of a story that constitutes the poetic solution to the [Augustinian] paradox of distension and intention. The fact that the story can be followed converts the paradox into a living dialectic [of episodic and configurational dimensions]. (I, 67)

Mimesis$_3$ marks the intersection of the world of the text and the world of the reader. The followability of a story, characteristic of mimesis$_2$, is actualized by reading it. The effect of a text on its reader is an integral component of the text's meaning. The text not only communicates its sense or meaning, but it also projects a world that can engage the reader in a creative "fusion of horizons."[4]

> What a reader receives is not just the sense of the work, but, through its sense, its reference, that is, the experience it

> brings to language and, in the last analysis, the world and the
> temporality it unfolds in the face of this experience. (I,
> 78–79)

Here, Ricoeur is relying on his earlier work, especially his debate
with structuralists on the referentiality of language (*The Rule of
Metaphor*). Both fictional and historical narratives have referential intentions and make truth claims. In fact, the two categories of
narrative have reciprocal tendencies, such that Ricoeur shall argue
for an *interweaving* reference between history and narrative
fiction. To make a narrative—any narrative—is to remake actions,
which resignifies a world that I might inhabit.

Part I of *Time and Narrative* ends with a forward glance, toward
Part IV, where the vast implications of the dialectic of time and
narrative shall become apparent in the difficult three-way conversation among history, literary criticism, and phenomenological
philosophy. This conversation receives its impetus from the referential implications of mimesis$_3$. Much rests upon a comparison of
Augustine's and Heidegger's understandings of the hierarchization
of our temporal experience. "Hierarchization" here refers to the
(Heideggerian) insight that we have different levels of temporal
experience and that the levels are graded according to how closely
they approximate a limited temporal experience. What limits our
temporal experience is ultimately the principle by which it can be
hierarchized (or the criterion that can distinguish its levels). In
Augustine's case, that limit is eternity: The hierarchization of
temporal experience reaches its upper limit in the approximation of
eternity by time. But in Heidegger's case, that limit is death: The
hierarchization of temporal experience is actualized to the extent
that Dasein is authentically being-toward-death. In the next two
chapters, which deal with Part IV of *Time and Narrative,* I shall
fully investigate Ricoeur's efforts to think about eternity and death
(Augustine and Heidegger) at the same time.

The Aporias of the Experience of Time:
Book XI of the *Confessions*

In Part I of *Time and Narrative,* Ricoeur attempts to think
together two rather disparate texts (*Confessions* and *Poetics*).

Augustine's *Confessions* give us a poignant sense of how difficult it is to think about something like time (or life) as such. How can we comprehend something so elusive? Surely, time is something and not nothing, but what exactly is it? Is it merely another name for "change"? Does it have extension in space? These are questions with which Augustine wrestles. His rather tentative answers are eked out of an aporetical style that can never find indisputable evidence for a position, except through a *via negativa* where all possible counterpositions are reduced to absurdity. Augustine's rhetoric plays right into Ricoeur's hands. Remember that Ricoeur, following and agreeing with Augustine, wants to insist that a *theoretical* resolution to the problem of time is impossible. But by coupling the *Confessions* with the *Poetics,* he demonstrates how a poetic or narrative response to the quandaries of time can be satisfactory. Aristotle's mimesis is the inverted image of Augustine's *intentio* such that concordance (understanding) can triumph over discordance (chaos).

Ricoeur is primarily interested in Book XI of the *Confessions.* It is commonplace to divide the *Confessions* into two parts (autobiography: Books I-X; biblical/theological reflection: Books XI-XIII) and to assume that the parts have little to do with each other. So there is nothing unusual about Ricoeur's isolation of Book XI from the rest of the text.

But Ricoeur further divides Book XI into two parts and then treats them separately.[5] The first part, for Ricoeur, begins in chapter 14 and ends in chapter 29 (more specifically, 14.7–29.38). It entails Augustine's discussion of time alone, apart from the overarching discussion of time in its relation to eternity (Ricoeur's "part two" of Book XI: chapters 1–14.6, 29.39–31.41). So, first, Ricoeur isolates Augustine's philosophical theory of time from its obvious religious context. He does so in order to demonstrate its aporetical character, which validates the thesis that any discussion of time is theoretically insoluble. Then, Ricoeur places Augustine's philosophical theory of time back into context, at least partially. Does this strategy violate Augustine's text? Ricoeur readily admits that it does. But he thinks that the discussion of eternity at the end of his own chapter rectifies the violation. Let us take a closer look at the *Confessions* in order to judge Ricoeur's interpretation. We can follow Ricoeur's own strategy by beginning with chapter 14 at the point where Augustine asks, "What, then, is time?"

Book XI: Part One

Ricoeur's treatment of chapters 14.7 through 29.38 (his "part one" of Book XI) is comprised of three parts that follow the natural progression of Augustine's argument. In Ricoeur's words,

> The notion of *distentio animi,* coupled with that of *intentio,* is only slowly and painfully sifted out from the major aporia with which Augustine is struggling, that of the measurement of time. This aporia itself, however, is inscribed within the circle of an aporia that is even more fundamental, that of the being or the nonbeing of time. (I, 7)

So in the first stage of his argument (chapters 14–20), Augustine wrestles with the question of whether or not time exists (the being or the nonbeing of time). Settling that issue, he is able to confront head-on the problem of measuring time (chapters 21–26). The reason Augustine had to settle the existence question first is simply that we cannot measure what does not exist. First we need to determine whether or not time exists, then we can worry about measuring it. The second stage of the argument concludes when Augustine determines that what we measure when we measure time is the extension of our own minds. This discovery leads to the final stage where the *intentio/distentio* dialectic is disclosed (chapters 27–29). We can take up the three stages of Augustine's argument in turn and also follow Ricoeur's interpretations of them.

Augustine begins by claiming that it is not easy to explain what time is, even though we readily understand what people mean when they speak of time.

> What, then, is time? There can be no quick and easy answer, for it is no simple matter even to understand what it is, let alone find words to explain it. Yet, in our conversation, no word is more familiarly used or more easily recognized than "time." We certainly understand what is meant by the word both when we use it ourselves and when we hear it used by others.[6]

In ordinary discourse, we speak of time and we refer to past, future, and present times. And we seem to understand what we mean by these terms. How can we interpret this fact (i.e., that we seem to know what we are talking about when we talk of time)? For Augustine, that we are able to speak meaningfully of time must mean that what we refer to exists in some sense. "I can confidently say that I know that if nothing passed, there would be no past time; if nothing were going to happen, there would be no future time; and if nothing *were,* there would be no present time" (ch. 14, 264). But in what sense, precisely, can we say that these three times exist? A skeptical voice arises in Augustine's query. How can the past and the future *be* when the past is no longer and the future is not yet? How can the present have any being either when what makes it distinctive is its passing away? Here, then, is the aporia of the being or nonbeing of time: We speak of it, therefore it must exist; yet upon examination, neither past nor future nor present seems to be anything at all.

In order to resolve this problem of time's existence, Augustine immediately refers to the fact that we also speak of periods of time that are either long or short. "Yet we speak of a 'long time' and a 'short time,' though only when we mean the past or the future" (ch. 15, 264). Again, that we speak meaningfully of time in terms of measurable periods must imply that such things as time periods exist. Augustine is assuming, of course, that ordinary discourse is trustworthy. (His theory of time is an interpretation of what we mean by speaking of time as actually existing.) A period of time is something that has duration. We can never say that the present has duration because it is merely the point of time when the future passes into the past. So only the past and the future have duration. The voice of skepticism enters again: How can we measure or assess the duration of something that does not exist (i.e., a period of past time or a period of future time)? We can measure only periods of time while they are passing because the future and the past can exist only in the moment of the present. "The conclusion is that we can be aware of time and measure it only while it is passing. Once it has passed it no longer is, and therefore cannot be measured" (ch. 16, 266). In order to exist, something must be visibly present (*praesens*); being implies presence (both spatial and temporal) for Augustine. Since the future and the past exist in some sense, they must be existing in the present.

> So wherever they are [the future and past] and whatever they are, it is only by being present that they *are.* . . . My own

childhood, which no longer exists, is in past time, which also
no longer exists. But when I remember those days and
describe them, it is in the present that I picture them to
myself, because their picture is still present in my memory.
. . . By whatever mysterious means it may be that the future
is foreseen, it is only possible to see something which exists;
and whatever exists is not future but present. (Ch. 18,
267–68)

Since, strictly speaking, only the present can exist, our ordinary
discourse about three times (future, past, present) needs correc-
tion. It might be more accurate, according to Augustine, to say
that the future is the present of future things, that the past is the
present of past things, and that the present is the present of present
things. At this juncture, the end of the first stage of his argument,
Augustine gives a tentative glimpse of his final solution to the
question of time. He suggests that these three times, or three
aspects of the present, exist only in the mind. "The present of past
things is the memory; the present of present things is direct
perception; and the present of future things is expectation [*praesens de praeteritis memoria, praesens de praesentibus contuitus,
praesens de futuris expectatio*]" (ch. 20, 269; 281). Thus ends the
first stage of his argument. It began with Augustine trusting our
ordinary use of language enough to skirt the skeptic's conclusion
that time cannot possibly exist, and it ends with a portention that
follows from a critical adjustment to our ordinary sense of time.
What do we mean by speaking of time as actually existing? That is
Augustine's basic question. A definitive answer is now starting to
take shape: In our ordinary understanding of it, time is a mental
phenomenon rather than a cosmic one.

Ricoeur thinks that this first stage of Augustine's argument
about the being and nonbeing of time is fragile and enigmatic.
Particularly enigmatic for Ricoeur is Augustine's reversion to
quasi-spatial language: "If the future and the past do exist, I want
to know *where* they are" (ch. 18, 267). As we saw, Augustine
eventually claimed that the future and past exist only in the mind.
But what is the basis for this turn to quasi-spatial language? Is it
because the question is posed in terms of "place" that such an
answer is given? "Or," asks Ricoeur, "is it not instead the
quasi-spatiality of the impression-image and the sign-image,

inscribed in the soul, that calls for the question of the location of the future and past things?" (I, 12). Even though it is too early in Augustine's argument to adjudicate between these lines of questioning, Ricoeur sides with the latter perspective since it anticipates the notion of *distentio animi,* which is crucial to Ricoeur's reading of Augustine.

In the second stage of Augustine's argument (chapters 21–26), he confronts the problem of measuring time. He first summarizes what he has learned in the initial stage of his inquiry: "We measure time while it is passing through the present since only as present can time exist to be measured." Yet when Augustine asks what standard we judge or measure a length of time against, he once again plunges into aporetical quandary.

> But to what period do we relate time when we measure it as passing? To the future, from which it comes? No: because we cannot measure what does not exist. To the present, through which it is passing? No: because we cannot measure what has no duration. To the past, then, towards which it is going? No again: because we cannot measure what no longer exists. (Ch. 21, 270)

Augustine takes a new tack in order to solve this puzzle concerning the means for measuring time. He asks if time is constituted by the movement of heavenly bodies such as the sun. As Augustine indicates, it is rather common to assume that time is nothing but the movement of the sun and the stars. But such a hypothesis Augustine rejects categorically. The motion of the sun does not constitute time; instead, it is *by* time that we measure the course of the sun.

> I cannot therefore accept the suggestion that time is constituted by the movement of heavenly bodies, because although the sun once stood still in answer to a man's [Joshua's] prayer, so that he could fight on until victory was his, the sun indeed stood still but time continued to pass. The battle went on for as long as was necessary and was then over. I see time, therefore, as an extension of some sort [*video igitur tempus quandam esse distentionem*]. (Ch. 23, 272; 285)

Even if the sun stops its journey, time continues. The Bible indicates as much in its account of Joshua's battle (see Josh. 10:13). So, according to Augustine, the movement of a body—whether it is the sun or a potter's wheel—is not the same as the means by which we measure the duration of that body's movement. But a period of time must be an extension (*distentio*) of something if it can take the measure of any movement. This is the crucial question for Augustine at the end of this stage of his argument: Of what is time an extension? When we measure the movement of bodies, we measure a period or an extension of time that corresponds to the movement. What is this extension? "I begin to wonder," says Augustine, "whether it is an extension of the mind itself" (ch. 26, 274).

Ricoeur dwells on Augustine's basic enigma in this second stage of the argument: Time has no extension in space, and yet we do measure it. How can we measure that which has no visible duration? "In order to resolve the enigma," states Ricoeur, "the cosmological solution must be rejected so that the investigation will be forced to search in the soul alone, and hence in the multiple structure of the threefold present, for the basis of extension and of measurement" (I, 14). So, first, Augustine investigates the "cosmological solution": Can time be equated with the movement of heavenly bodies? If so, then the problem of what to measure is solved—just measure time as the extent of the movement of the sun. But Augustine rejects that possibility. He claims that time is the basis for measuring any movement but is not identical to that movement. What possibilities are left then? If time is a measurable extension, and if the movement of the sun is not coextensive with time, then time must be an extension of the mind itself. According to Ricoeur, this illustrates the aporetical nature of Augustine's rhetoric to the extent that there is no real phenomenological core to Augustine's argument. How does Augustine arrive at his insight that time is an extension of the soul? He does it by reducing other hypotheses to absurdity. Ricoeur explains:

> If there is a phenomenological core to this assertion [that time is an extension of the soul], it is inseparable from the *reductio ad absurdum* that eliminated the other hypotheses: since I measure the movement of a body by time and not the other way around—since a long time can only be measured

by a short time—and since no physical movement offers a
fixed unit of measurement for comparison, the movement of
the stars being assumed to be variable—it *remains that* the
extension of time is a distension of the soul. (I, 15–16)

In the third and final stage of his argument about the nature of
time (chapters 27–29), Augustine connects the themes of the two
prior stages. In the first stage, the idea of a threefold present was
proffered in order to resolve the enigma of a being (time) that
lacks being. In the second stage, the idea of a distension of the
mind was proffered in order to resolve the enigma of the extension
of a thing (time) that has no extension. "What remains, then,"
according to Ricoeur, "is to conceive of the threefold present *as*
distension and distension *as* the distension of the threefold
present. This is the stroke of genius of Book XI of Augustine's
Confessions, in whose wake will follow Husserl, Heidegger, and
Merleau-Ponty" (I, 16).

In chapter 27, Augustine finally finds the proper way to
understand what it is we measure when we measure time. He
begins by analyzing three different kinds of sound—one that has
just ended, one that is continuous, and one that entails long and
short syllables. All three examples converge upon one conclusion:
When we measure time we measure the impression that things
leave on our minds.

In its fullest declaration, Augustine's theory is that the mind
performs three functions—expectation, attention, and memory—
such that "the future, which it expects, passes through the present,
to which it attends, into the past, which it remembers" (ch. 28,
277). Augustine illustrates his intact theory by analyzing the
recitation of a psalm (Ricoeur calls it the "crown jewel" of the
Confessions).

> Suppose that I am going to recite a psalm that I know. Before
> I begin, my faculty of expectation is engaged by the whole of
> it. But once I have begun, as much of the psalm as I have
> removed from the province of expectation and relegated to
> the past now engages my memory, and the scope of the
> action which I am performing is divided between the two
> faculties of memory and expectation, the one looking back
> to the part which I have already recited, the other looking
> forward to the part which I have still to recite. But my faculty

> of attention is present all the while, and through it passes
> what was the future in the process of becoming the past. As
> the process continues, the province of memory is extended
> in proportion as that of expectation is reduced, until the
> whole of my expectation is absorbed. This happens when I
> have finished my recitation and it has all passed into the
> province of memory. (Ch. 28, 278)

Ricoeur thinks that this analysis of a recitation marks the point
at which the theory of *distentio* is joined to that of the threefold
present (I, 19–20). However, the threefold present in this context
(stage three of Augustine's argument) is not the same as it was in
the earlier stage where Augustine was attempting to say how three
different times can exist (i.e., as three different aspects of the
present). Previously, the present was construed in a passive sense:
It is merely a transition point, a point of passage from the future
(which does not exist) to the past (which also does not exist).
Now, in stage three, the present is given an active interpretation.
The mind's attention (*intentio praesens*), which is the present
component of "mental time," actively relegates expectations into
memories. So the theory of *distentio* (what we measure are
impressions that indicate extension in the mind) is joined to that of
the threefold present (time can only exist as the present) when
expectation and memory are understood both as extensions of the
mind and as activities of mental attention.

> It is *in* the soul, hence as an impression, that expectation and
> memory possess extension. But the impression is in the soul
> only inasmuch as the mind *acts,* that is, expects, attends, and
> remembers. (I, 19)

Hence, says Ricoeur, "the theory of the threefold present, refor-
mulated in terms of the threefold intention, makes the *distentio*
arise out of the *intentio* that has burst asunder" (I, 20). Here, then,
is the *intentio/distentio* dialectic that Ricoeur wants to pair with
Aristotle's *mimesis/muthos* dialectic.

The supreme enigma of Augustine's theory of time, according
to Ricoeur, is that the soul "distends" itself as it engages itself:
The more the mind makes itself *intentio,* the more it suffers
distentio. What does all this mean, and what is its significance?
That the mind or soul is distended indicates that it is divided

against itself. There is never a complete coincidence of the mind with itself because it is torn between the past and the future. Attention to the past is *different* than attention to the future. The two can never coincide; the soul can never be at rest since it is forever having to relegate one perspective to the other. As temporal phenomenon, the soul is divided; it is discordant with itself. In order to achieve understanding or concordance, it must continually attend to the split or chasm that it straddles. For Augustine, there is only one intention that can save the soul from its own distension.

> I am divided between time gone by and time to come, and its course is a mystery to me. My thoughts, the intimate life of my soul, are torn this way and that in the havoc of change. And so it will be until I am purified and melted by the fire of your love and fused into one with you. . . . Then I shall listen to the sound of your praises and gaze at your beauty ever present, never future, never past. . . . I am intent upon this one purpose. (Ch. 29, 278–79)

Only within the eternal—the ever present—can temporal distension be overcome. The task of life, for Augustine, is to intend the eternal: to seek (and see) God.

Augustine reduced the extension of time to the distension of the soul. He tied this distension to the slippage or noncoincidence among the three aspects of the present. Thus, discordance emerges "again and again out of the very concordance of the intentions of expectation, attention, and memory" (*Time and Narrative* I, 21). Ricoeur responds:

> It is to this enigma of the speculation on time that the poetic act of emplotment replies. But Aristotle's *Poetics* does not resolve the enigma on the speculative level. It does not really resolve it at all. It puts it to work—poetically—by producing an inverted figure of discordance and concordance. For this new solution, Augustine does leave us one word of encouragement. The fragile example of the *canticus* recited by heart suddenly becomes, toward the end of the inquiry, a powerful paradigm for other *actiones* in which, through engaging itself, the soul suffers distension: "What is true of the whole psalm is also true of all its parts and each syllable.

It is true of any longer action in which I may be engaged and
of which the recitation of the psalm may only be a small part.
It is true of a man's whole life, of which all his actions are
parts. It is true of the whole history of mankind, of which
each man's life is a part" [Book XI, ch. 28.38]. The entire
province of narrative is laid out here in its potentiality, from
the simple poem, to the story of an entire life, to universal
history. It is with these extrapolations, which are simply
suggested here, that the present work is concerned. (I,
21–22)

In germ, Ricoeur's entire project is anticipated by Augustine
himself. What is true of the psalm is true of life and history.
Understanding is achieved through a process of relegation (narration) that manifests a coherent whole (concordance) despite the
finitude of one's attentive powers (discordance).

Book XI: Part Two
In the final section of his chapter on the *Confessions,* Ricoeur
attempts to place the analysis of Augustine's theory of time back
into context. He cannot overlook the fact that chapters 14 through
29 ("part one" of Book XI) are framed by a meditation on eternity.
So what happens to the *distentio/intentio* dialectic when it is
viewed from the perspective of eternity? Ricoeur thinks that the
meditation on eternity affects the speculation concerning time in
three major ways. Before discussing Ricouer's remarks, let me
introduce the reader to Augustine's argument on eternity in
chapters 1 through 13 and 29 through 31.
    In chapter 3, Augustine asks God to help him understand
Genesis 1.1 ("In the beginning God made heaven and earth").
That appeal sets the agenda for the entire book. The main theme is
to compare Creator and creature. Eternity and time are principal
figures in such a comparison. Apart from the first four chapters,
which are introductory, there are two different sections in this
"part" of Book XI. The first section (chapters 5–9) deals with the
question of how God created the heavens and the earth. The
second section (chapters 10–13, 29–31) deals with the question of
what God was doing prior to creation, and, more broadly, with the
question of guidance for the unenlightened.
    "But by what means did you make heaven and earth?" asks

Augustine at the beginning of chapter 5. "What tool did you use for this vast work? You did not work as a human craftsman does, making one thing out of something else as his mind directs" (ch. 5, 257). Augustine is saying that if God is the creator of the universe, then the universe could not have existed as something for God to use in crafting it. But if nothing was available for God's use, how was the universe created? He concludes that God creates through his Word (*Verbum*) alone. That conclusion leads to another problem, however. In chapter 6, Augustine asks God how he is able to speak. Is God's speech subject to the laws of time? That is, do God's syllables sound and then die away like ours do? No, says Augustine. God's Word is silent and eternal and is "heard" inwardly by the intelligence or mind. Otherwise, the creative Word would depend upon time and other features of the universe that the Word is supposed to have created.

> In your Word all is uttered at one and the same time, yet eternally. If it were not so, your Word would be subject to time and change, and therefore would be neither truly eternal nor truly immortal. (Ch. 7, 259)

Having established that God creates by Word alone, and that God's Word is co-eternal with God's own self, Augustine faces another problem. Now he wonders why the things that God creates do not come into being all at once. All one can say, states Augustine, is that "whatever begins to be, or ceases to be, does so at the moment when the eternal reason knows that it should begin to be or cease to be, although in the eternal reason there is no beginning or ending" (ch. 8, 259). God's Word, or the Beginning, is eternal reason, as is Christ. Things get complicated and interesting when Augustine brings the incarnate Word into the picture. Suddenly, there is talk of restoration and return—redemption—and not just creation. The ongoing nature of creation, such that God determines the proper time for things to begin and cease, implies an overarching purpose that includes a notion of redemption.

> Even when we learn from created things, which are subject to change, we are led to the Truth which does not change. And there we truly learn, as we stand by and listen to him

and rejoice at hearing the bridegroom's voice, restoring
ourselves to him who gave us our being. He [Word/Christ] is
therefore the Beginning, the abiding Principle, for unless he
remained when we wandered in error, there would be none to
whom we could return and restore ourselves. But when we
return from error, we return by knowing the Truth; and in
order that we may know the Truth, he teaches us, because he
is the Beginning and he also speaks to us. (Ch. 8, 260)

God creates the universe, sending it forth through his creative
Word. Yet part of the plan of creation, as Augustine sees it, is for
the redemptive Word (Christ) to draw everything back into God.
There is an initial going forth and a subsequent return. Redemption is a kind of reverse or second creation: Everything must
return to the eternal source, the Beginning. Both movements—
creation and return—are orchestrated by the Eternal Word. This
double duty can be confusing. But it is one way Augustine and
other theologians were able to graft a Christian (and particularly a
Johannian) perspective onto the Genesis creation account. The
interesting question is this: Does Augustine's use of the term
"time" entail a double sense that follows from the double sense in
his use of the term "Word"? I think that it does.

According to my reading of the *Confessions,* Augustine uses
"time" in two different senses, depending upon the perspective he
has in mind—beginnings or endings. With respect to beginnings,
or creation as such, time is a special part of the universe that God
makes. "You are the Maker of all time," says Augustine (ch. 13,
263). The universe, and everything in it, is subject to the laws of
time (i.e., the laws of change). Time elapses or flows, and we must
suffer it ("Time could not elapse before you made it" [ch. 13,
263].) This elapse of time as universal law or principle of change
is very different from the "mental time" of Book XI, part one. As
explained previously, mental time is something that we partially
control rather than suffer; it is the attentive relegation of the future
into the past. With this latter sense of time, the only one that
Ricoeur credits to Augustine, the perspective is one of endings
rather than beginnings.[7] The time of endings pertains to redemption rather than creation. Here the Word, as Christ or Truth,
prompts our mind and memory in order to guide us back to the
Eternal. In this sense, time itself is actually a grace that we

cooperate with; it is an extension of the Word itself rather than a created effect. As Augustine says,

> The past is always driven on by the future, the future always follows on the heels of the past, and both the past and the future have their beginning and their end in the eternal present. If only men's minds could be seized and held still! They would see how eternity, in which there is neither past nor future, determines both past and future time. (Ch. 11, 262)

In chapter 10, Augustine indicates that those who ask what God was doing before he made heaven and earth are steeped in error. They wonder how a new will or idea—to create the universe—could interrupt the eternal repose of God. As Augustine sees it, the problem is that they have misunderstood eternity. Questions of future and past, or before and after, do not apply to eternity. The future and past are aspects of time, and Augustine says that time and eternity are not comparable (ch. 11). Time is never still, eternity is for ever still; both past and future are determined by eternity. Augustine is not afraid to put things bluntly when need be. "Before he made heaven and earth, God made nothing" (ch. 12, 262). He advises people to not let their misconceptions of time or eternity lead them astray. Think more carefully, he counsels in chapter 13. Since God made all time, God was before all time, "and the 'time,' if such we may call it, when there was no time was not time at all" (ch. 13, 263).

In chapter 30, after his sustained speculation on the nature of time (chapters 14–29), Augustine returns to the problem of what to tell those who ask what God was doing before he made heaven and earth. This concern to advise others links the last chapters of Book XI with the earlier chapters. "Let them have done with this nonsense," he admonishes (ch. 30, 279). "Let them instead be intent on what lies before them." Augustine seems to be suggesting that we not get too hung up on how everything in the universe works (this after [and as a result of] his own Herculean efforts to understand time!). Rather, we should focus on things that lie somewhat under our control—the redemption of our souls. To tell us to be *intent* on what lies ahead, given the prior discussion of what an intention is, means that we should try to heal the rift in our souls by imitating the Eternal (or by becoming more like God). Only in eternity can past and future, beginning and ending, be brought together. Yet in life we are torn, distended, between beginning and ending. By intending the eternal, we can ameliorate that predicament.[8]

In the last section of his chapter on the *Confessions*, Ricoeur attempts to mitigate the fact that he isolated Augustine's psychological theory of time from its context within a meditation on eternity. He still insists that the sufficient sense of the *distentio animi* can be grasped outside of the contrast with eternity. "And yet," he says, "something is missing from the full sense of *distentio animi,* which the contrast with eternity alone can provide" (I, 22). He sees three ways the meditation on eternity affects the speculation on time. First, eternity functions as a limiting idea for all speculation about time. Second, it intensifies the experience of *distentio* on the existential level. Third, it calls upon the experience of *distentio* to surpass itself by imitating the Eternal. As I interpret it, Ricoeur's reading of part two of Book XI in terms of three types of interaction between eternity and time follows the creation/redemption (beginning/ending) scheme that I discussed above. Hence too my suggestion that Ricoeur can only partially ignore the theological framework of Augustine's argument. The distinction or contrast is between the first and third ways in which eternity affects time. The second way bridges the difference. Let me explain.

First of all, eternity functions as a limit-idea with respect to time. Here, the full force of the contrast between eternity and time is seen. The context is the ontological difference between Creator and creature. Compared to God, the human is as nothing. Compared to eternity, time is struck with negativity; it is ontologically lacking. For example, Augustine says that eternity is forever still while time is never still. Or, eternity is ever present, but time is never all present at once. Eternity as limit-idea is not devoid of an experience. It is no mere thought lacking an object. Again, the context is the difference between Creator and creature. Augustine addresses himself to the Eternal as to the Creator of heaven and earth. He compares himself to the Almighty. "It is the recoil effect of this 'comparison' on the living experience of the *distentio animi* that makes the thought of eternity the limiting idea against the horizon of which the experience of the *distentio animi* receives, on the ontological level, the negative mark of a lack or a defect in being" (I, 26). The absence of eternity, for Augustine, is a real lack that is sorrowfully felt. This felt lack, when intensified to the level of a lamentation, becomes the second way in which eternity affects the theory of time. It marks the most intense realization of the ontological difference separating creature and Creator. It also marks the turning of the deprived and torn soul toward that which it lacks, that is, the stillness and peace of the eternal present. This leads to the third way in which the dialectic of time and eternity affects the theory of *distentio.*

The third way, whereby eternity produces a hierarchy of levels of temporalization by providing a model for time to imitate, is the opposite of the first way. Rather than focusing on creation and the ontological difference between Creator and creature, the third way focuses on redemption and the resemblance between eternity and time. "This resemblance is expressed in time's capacity to approximate eternity" (I, 28). By listening to the teachings of the inner Word, we can return to God, our Beginning.

> Between the eternal *Verbum* and the human *vox* there is not only difference and distance but the relation of teaching and communication. . . . The teaching, we could say, bridges the abyss that opens up between the eternal *Verbum* and the temporal *vox*. It elevates time, moving it in the direction of eternity. (I, 29)

The elevation of time by eternity is the return of redemption.

Summing up his anchorage of the dialectic of *intentio* and *distentio* in that of eternity and time, Ricoeur focuses on a key point. He insists that one definite advantage is gained by bringing eternity into the discussion of time. Eternity is able to deepen our sense and experience of time.

> Indeed, it was necessary to confess what was other than time [eternity] in order to be in a position to give full justice to human temporality and to propose not to abolish it but to probe deeper into it, to hierarchize it, and to unfold it following levels of temporalization that are less and less "distended" and more and more "held firmly," *non secundum distentionem sed secundum intentionem* [ch. 29.39]. (I, 30)

So Ricoeur has found his *philosophical* way to retrieve Augustine's sense of eternity. It offers us a horizon or limit to shoot for, a vast perspective on our own precarious situation. In order to better understand time, imagine eternity—that is Ricoeur's Augustinian suggestion. Eternity functions as a regulative idea (à la Kant) in Ricoeur's theory on narrative and time. (I shall explain this point in more detail in the next two chapters.)

But what of the *religious* sense of eternity in Augustine's text? Ricoeur is silent on this issue (although he finally broaches the topic in his "Conclusion," as we shall see later). I have criticized his psychological reading of Augustine's theory, claiming that it would be better to call it a "redemptive theory" of time, in order to keep the religious and theological issues in the forefront of the analysis of the *Confessions.* In chapter 4, I shall argue that redemptive time, in Augustine's own (unthought or unspoken) understanding, is actually the being of God when God is being other than God (as Eternal Word).

Notes

1. Ricoeur's use of the term "aporia" is comparable to Kant's use of the term "antinomy." An aporia, like an antinomy of reason, indicates a critical point where it becomes philosophically impossible to resolve the discrepancy between two alternative points of view. The aporia, in Augustine's case, is the unresolvable conflict between the soul's attention to the past and its attention to the future. The soul can never attend to both at the same time. Rather, it is hopelessly distended between their different perspectives.
2. There is an obvious correspondence between Ricoeur's threefold mimesis and the "hermeneutical arch" that he developed in such writings as *Interpretation Theory*. The arch describes the three basic moments of any interpretation. First, an interpreter formulates a preunderstanding of a text. Next, he or she subjects the preunderstanding to critical testing. Finally, the critically approved explanation is appropriated into the interpreter's life-world. The pattern is roughly the same in mimesis: From a prenarratival understanding of experience, one moves to a narrated understanding that ties the episodes of life together, until finally one's basic experiences of life have been transformed by narrative structures. See chapter 4 herein for a further discussion of Ricoeur's hermeneutical theory.
3. See chapter 3 for an extensive discussion of Heidegger.
4. "Fusion of horizons" is a phrase coined by Hans-Georg Gadamer (see *Truth and Method*, 273ff.). Gadamer claims that the process of understanding is one of mutual exchanges wherein two people (or two epochs) forge a broad, shared perspective. Understanding, then, entails a fusion of horizons. It is an apt phrase to use in describing Ricoeur's hermeneutics as well since he has aligned himself with Gadamer. Ricoeur develops the methodological side of Gadamer's (and Heidegger's) insight into the understanding process. The moment of appropriation in Ricoeur's hermeneutical arch (see chapter 4 for a full discussion of Ricoeur's hermeneutical theory)—where the worlds of the text and the reader interact—is the point in Ricoeur's work where the notion of "fusion of horizons" is applied.
5. In a footnote (*Time and Narrative*, vol. 1, Part 235, note 30), Ricoeur

indicates the rationale for his division of Book XI. He explains that
the editor of the *Confessions* in the "Bibliotheque Augustinienne"
divided the book this way: I. The creation and the creating Word
(3.5–10.12); II. The problem of time: (a) before the creation (10.12–
14.17), (b) the being of time and its measurement (14.17–29.39).
Ricoeur simplifies this scheme by combining I and IIa.
6. When citing Book XI hereafter, I shall give the chapter number and
then the page numbers from the 1961 Penguin edition. In this case,
the quote is from chapter 14, pages 263–64. When I include
references to the Latin original, the page numbers from the 1981
Teubner edition shall follow the numbers from the English text.
7. This criticism will have some bearing on the next chapter where I
discuss Ricoeur's contrast between Aristotle's cosmological theory
of time and Augustine's psychological theory. It seems to me that
Ricoeur has simplified Augustine's position in order to make it fit his
own theoretical schema. What he overlooks is the theological framework of Augustine's discussion of time. The creation/redemption
theme in Augustine's text seems able to embrace and presage
Ricoeur's basic argument. In a real sense, *Time and Narrative* is an
extended commentary on the *Confessions*—perhaps more than Ricoeur realizes.
8. Between creation and redemption is the story of a pilgrim's life.
Ricoeur's theory of narrative, which is suspended between cosmological and phenomenological theories of time (see my next chapter), is
no doubt an interpretation of this traditional Christian/Augustinian
philosophy of life.

Chapter 2

From Aristotle and Augustine to Heidegger: Reflecting on Time in the West

Ricoeur's efforts to think death and eternity together, or to confront Heidegger with Augustine, are developed in Part IV of *Time and Narrative*. In Part IV, Ricoeur interprets the tradition of philosophical reflections on time in terms of the basic aporia or split between a phenomenological perspective and a cosmological perspective. He traces the phenomenological (or subjectivist) strand of reflections from Augustine to Husserl to Heidegger. That strand of the tradition is contrasted with the cosmological reflections that stem from Aristotle (*Physics*) to Kant to contemporary science. Then Ricoeur argues for a narrative perspective on time that can mediate the aporia entrenched in our thinking about time. It is within the context of this narrative perspective that Ricoeur brings death and eternity together. They constitute the limiting horizons for the historical and fictional aspects of narrative consciousness, which stem the difference between cosmic time and phenomenological time.

For Ricoeur, what links *Being and Time* to the *Confessions,* despite a separation of fifteen centuries, is a common effort to think the meaning of time phenomenologically. He argues that Heidegger's hermeneutic phenomenology participates in and extends the tradition of phenomenological investigation into the nature of time begun by Augustine. In this chapter, I review the steps Ricoeur takes in moving from Augustine to Heidegger and offer original interpretations of the texts he singles out. An analysis of *Being and Time,* and a more thorough articulation of Ricoeur's confrontation between Heidegger and Augustine, is presented in chapter 3.

Time and Narrative, Part IV: Narrated Time

"This fourth part of *Time and Narrative,*" says Ricoeur, "is aimed at as complete an explication as possible of the hypothesis that governs our inquiry, namely, that the effort of thinking which is at work in every narrative configuration is completed in a refiguration of temporal experience" (III, 3). The refiguration of our temporal experience is the final outcome of the mimetic process. Hence, this part of the text continues the discussion of mimesis$_3$ begun at the end of Part I. In a sense, one could argue that Parts II and III actually interrupted the continuity of the text. They offered a "detour" that extended the discussion of Ricoeur's narrative model (mimesis$_2$) to the two different fields of narrative investigation—history (Part II) and fiction (Part III).[1] This detour provided the necessary supplementation for a fuller understanding of poetic refigurations of our temporal experience, which involve the "interweaving" of historical and fictional powers of reference.

The two sections of Part IV further develop Ricoeur's initial polarity between Augustine and Aristotle (Part I). The general thesis that served as the basis for that initial contrast still holds here: Narrative composition constitutes an answer to the aporetic character of speculations on time. The first section, following the Augustinian side of the contrast, presents an aporetics of temporality as that which stands over against the power of narrative to refigure our temporal experience. The idea that there can never be a phenomenology of temporality free of every aporia, first introduced in the chapter on Augustine's *Confessions,* is demonstrated further in discussions of Aristotle (*Physics*), Kant, Husserl, and Heidegger. The second section, following the Aristotelian (*Poetics*) side of the initial confrontation, further examines the creative resources by which narrative activity—whether historical or fictional—responds to and corresponds with the aporetics of temporality.

Section One: The Aporetics of Temporality

Augustine, as we saw, tried to derive the principle of the extension and measurement of time from the distension of the mind alone. But Augustine's exclusive focus on a psychological conception of time (in chapters 14–29), according to Ricoeur, cannot replace Aristotle's cosmological conception:

> Behind Aristotle stands an entire cosmological tradition, according to which time surrounds us, envelops us, and dominates us, without the soul having the power to produce it. I am convinced that the dialectic of *intentio* and *distentio animi* is powerless to produce this imperious character of time and that, paradoxically, it helps conceal it. (III, 12)

There is, then, an unresolvable disagreement, an aporia, involving these two different conceptions of time, and Ricoeur's theory of narrative responds precisely to this aporia. The aporia between the psychological and cosmological conceptions of time is the basis for all other aporias in our attempts to think time.

Ricoeur maintains that Augustine never refuted Aristotle. The psychological (or phenomenological) theory of time is no substitute for the cosmological theory but, rather, must be added to it. Neither theory can be completely derived from the other. They constitute a fundamental duality that delimits the basic problem of time. Both sides of this aporia are necessary, yet a reconciliation is demanded. Narrative both confirms the aporia and brings it to poetic resolution.

According to Ricoeur, the confrontation between Augustine and Aristotle on time does not exhaust the whole aporetics of time. He extends and clarifies the debate between the phenomenological and cosmological senses of time by turning to Husserl and Kant, then to Heidegger. In his *Phenomenology of Internal Time-Consciousness,* Husserl attempts to make time itself appear. Such a feat, if possible, would free phenomenology of all the aporias we have come to expect from Augustine's discussion. But is a pure phenomenology of time possible? Can Augustine's uncertainties be cleared up by further analysis of time-consciousness? Ricoeur argues that Husserl's ambition to make time as such appear runs up against Kant's thesis of the objectivity of time (*Critique of Pure Reason*) that extends Aristotle's discussion of physical time. "For Kant, objective time—the new figure of physical time in a transcendental philosophy—never appears as such but always remains a presupposition" (III, 23). As presuppositional, objective time, the time implied in the determination of objects, is essentially invisible and can never be made to appear as such. As Augustine was to Aristotle in the initial confrontation between the two perspectives on time, Husserl is to Kant in the first extension of the discussion of the aporetics of temporality. In the second extension, Heidegger's "hermeneutic phenomenology" is opposed to our ordinary conception of time; Heidegger is to ordinary time what Augustine and Husserl are to Aristotle and Kant.

So the confrontation of Husserl and Kant leads to an impasse similar to the one in the confrontation of Augustine and Aristotle. Neither approach (phenomenological or cosmological) is sufficient in itself to explain time. Each side seems to exclude the other, and yet there are implied mutual borrowings. Ricoeur explains:

> On the one hand, we can enter the Husserlian problematic only by bracketing the Kantian problematic; a phenomenology of time can be articulated only by borrowing from objective time, which, in its principal determinations, remains a Kantian time. On the other hand, we can enter the Kantian problematic only on the condition of abstaining from all recourse to any inner sense that would reintroduce an ontology of the soul, which the distinction between phenomenon and thing in itself has bracketed. Yet the determinations by which time is distinguished from a mere magnitude must themselves be based on an implicit phenomenology, whose empty place is evident in every step of the transcendental argument. In this way, phenomenology and critical thought borrow from each other only on the condition of mutually excluding each other. We cannot look at both sides of a single coin at the same time. (III, 57)

Only through poetic construction can this aporia of two times, the time of the soul and the time of nature, be productively resolved.

Heidegger's hermeneutic phenomenology is different from Husserl's transcendental phenomenology because it entails the suspicion that what is closest to us in the sense of being most primordial—for example, our own time-consciousness—is inevitably concealed by our own inauthentic tendencies. Hence, a "hermeneutics" is required in order to uncover through understanding what always escapes direct appearance due to concealment. Our ordinary sense of time, most often associated with clock time, represents the cosmic perspective against which Heidegger's phenomenological perspective is construed. Again, Ricoeur makes the same point, though this time the aporia has reached its greatest pitch (confirming the thesis that the more advanced our phenomenology of time becomes, the more aporetical it becomes): The phenomenological and cosmic perspectives both mutually occlude and mutually depend upon each other, thus they constitute an inevitable aporia that is reflectively unresolvable, though not poetically unresolvable.

In effect, Ricoeur's discussion of the classic texts on time is a narrative built around the initial *agon,* or conflict, between Augustine and Aristotle. The conflict keeps repeating itself in other thinkers (Kant, Husserl, Heidegger), growing in intensity. Ricoeur's narrative (as opposed to his reflections on narrative) seems to instantiate that of which it speaks; in other words, his narrative illustrates his thesis that only through narrative can the theoretical impasse between the two approaches to time (phenomenological and cosmological) be overcome.

Section Two: Poetics of Narrative: History, Fiction, Time

Ricoeur argues that both history and fiction respond creatively to the aporias that become apparent in the phenomenology of time, principally the rift between lived (phenomenological) time and cosmic time. Lived time is reinscribed on cosmic time through the mediation of a third time, called historical, human, or narrated time. Each narrative type, history and fiction, makes its own unique contribution to this invented, mediating time. Ricoeur calls the relation between the two their "interweaving reference" or "interweaving refiguration." Both history and fiction respond to the same phenomenological quandaries, contributing different features to an overall narrative solution given in the intersection (interweaving) or mutual cooperation of the two types of narrative. The refiguration of time/praxis through narrative is accomplished finally by the conjoined efforts of history and fiction to have a transforming effect upon readers. Every narrative presents a world that is potentially inhabitable to a reader. The third, mediating time that is able to overcome the disjunction of the two times given by phenomenology is created by the mutual borrowing implied in the two modes of narrative. On this mutual intimacy, Ricoeur says that

> historical intentionality only becomes effective by incorporating into its intended object the resources of fictionaliza-

tion stemming from the narrative form of imagination, while the intentionality of fiction produces its effects of detecting and transforming acting and suffering only by symmetrically assuming the resources of historicization presented it by attempts to reconstruct the actual past. From these intimate exchanges between the historicization of the fictional narrative and fictionalization of the historical narrative is born what we call human time, which is nothing other than narrated time. (III, 101–2)

Of particular importance for this study is the discussion of time, eternity, and death in the fifth chapter of Part IV ("Fiction and Its Imaginative Variations on Time"), which extends Ricoeur's earlier work on fiction in Part III. In particular, two concepts, "fictive experience of time" and "imaginative variations," are carried over from Part III. The fictive experience of time is the temporal aspect of a reader's ability to inhabit the world projected by a literary text. The text's world is what opens the text to the outside, enabling it to escape its own self-enclosure. The world of the text constitutes a transcendence immanent in the text. Ricoeur developed this notion of fictive experience, wherein an imaginary world intersects with the reader's actual world, through the interpretation of three novels, each a tale about time: *Mrs. Dalloway* by Virginia Woolf, *Der Zauberberg* by Thomas Mann, and *À la recherche du temps perdu* by Marcel Proust. The point Ricoeur made in reading the three novels is that, through imaginative variations of plot and character, fictional works are able to explore the limits of our temporal experience and to explore the relations between time and eternity and eternity and death.

Now, in Part IV, Ricoeur wants to show that the imaginative variations of fiction contrast with the invariant reinscriptions of phenomenological time on cosmic time accomplished by history. The variable constitution of fictive time and the invariable constitution of historical time provide different responses to the aporias of temporality. The fictive narrator, unlike the historian, does not have to worry about reinscribing lived time upon cosmic time: "The time of fictional narrative has been freed from the constraints requiring it to be referred back to the time of the universe" (III, 128). This freedom enables fiction to explore the hidden resources of phenomenological time. "Fiction," to Ricoeur, "is a treasure trove of imaginative variations applied to the theme of phenomenological time and its aporias" (III, 128). It is finally in their manner of relating to the split opened up by reflective thinking between phenomenological time and cosmic time that fiction and history differ. Fiction stands closer to phenomenological time, history to cosmic time; their interweaving bridges the gap between the phenomenological and the cosmological.

Ricoeur argues that fiction liberates phenomenological time from the constraints of historical time by extending any phenomenology of time into an exploration of its upper limits. (Fiction also entails the liberating capacity to explore the boundaries of the mythical language that undergirds all philosophical inquiry.) The exploration of upper limits is as open to the Augustinian line of thinking about eternity as it is to the Heideggerian line of thinking about death. Here, then, is where Ricoeur attempts the confrontation between Augustine and Heidegger that he promised at the end of Part I. Everything hinges upon their different ways of conceiving the hierarchization of our temporal experience. For Augustine, the principle by which our temporal experience is hierarchized is given in the limit-experience of the eternal; for Heidegger, that principle is given in the experience of being-toward-death. Let me give an extended quotation, and then make some interpretive remarks. (I remind the reader that this is a preliminary discussion; I investigate Ricoeur's confrontation of Heidegger and Augustine more carefully in my next chapter.)

> For Augustine, this upper limit [to the process of temporalization] is eternity. And for the current in Christian tradition that incorporated the teachings of Neoplatonism, time's approximation of eternity lies in the stability of a soul at rest. Neither Husserlian phenomenology nor the Heideggerian hermeneutic of Dasein has continued this line of thinking. . . . As for *Being and Time,* its philosophy of finitude seems to substitute thinking about Being-towards-death for meditating on eternity. . . . Is this disjunction [between eternity and death] only apparent [or is it real]?
>
> The answer to this question can be sought on several levels. On the properly theological level, it is not certain that the conception of eternity is summed up in the idea of rest. We will not discuss here the Christian alternatives to the equating of eternity with rest. But on the formal level of a

philosophical anthropology—the level where Heidegger still situates himself in the period of *Being and Time*—it is possible to distinguish between the existential and the existentiell components in the pair that constitutes Being-towards-death and anticipatory resoluteness in the face of death. The function of attestation acribed to the latter with respect to the existential "Being-towards-death" allows us to think that this existential of universal mortality leaves open a vast range of existentiell responses, including the quasi-Stoic resoluteness affirmed by the author of *Being and Time*. For my part, I have unhesitatingly held mortality to be a universal feature of the human condition. . . . But I left hanging the question whether the existential component of Being-towards-death, and perhaps even existentiell modalities other than the Stoic tone given by Heidegger to resolution, including the modalities of Christian hope stemming in one way or another from faith in the Resurrection. It is in this interval between the existential and existentiell that a meditation on eternity and on death can be conducted. (III, 135–36)

Upon a first reading of this passage, I was convinced that Ricoeur brings Heidegger and Augustine together in much the same way that Bultmann brought Heidegger and Gogarten together in "The Historicity of Man and Faith."[2] The same principle seemed to be at work in both cases. Heidegger's distinction between the existential and the existentiell (or the ontological and the ontical) seemed able to account for the difference between death and eternity much as it was able, for Bultmann, to account for the difference between resolution and love (or philosophy and theology). For Heidegger, as I show more clearly in my next chapter, *existential* questions are concerned with what constitutes existence, or with the ontological structure of existence, while *existentiell* questions have an immediate or ontical concern for what is factically given in existence. Ontologies, in that they are concerned with the conditions of existence, provide the foundation for ontical sciences such as theology that are concerned with specific manifestations of existence. Following Heidegger's own view of the matter, Bultmann argues that philosophy is the ontological science that demonstrates the possibilities within which theology, as an ontical science, must operate. Hence, Christian love, according to Bultmann, is the authentic ontical actualization of human resolve understood ontologically. Upon a second reading, however, I was persuaded that nothing of this sort occurs in Ricoeur's confrontation of Heidegger and Augustine.

The last two sentences in the above passage only seem to suggest a Bultmannian synthesis. The key difference is that Ricoeur is critical of Heidegger's philosophical anthropology, whereas Bultmann accepted it as an ontological norm. Notice that Ricoeur says Christian faith is a potential existentiell modality *other than* Heidegger's resolve. Bultmann accepted that Heidegger's resolve is an ontological category that need never compete with an ontical faith. But Ricoeur is suggesting that Heidegger's existential analytic is fundamentally confused on this precise issue of the difference between the existential and the existentiell (or the ontological and the ontical).

In his treatment of Heidegger's hermeneutical phenomenology of time, Ricoeur argues that Heidegger confuses the existential and the existentiell due to a basic complicity between that distinction and the distinction between the authentic and the inauthentic.[3] Ricoeur asks: How can Heidegger maintain the distinction between the existential and the existentiell while at the same time insisting that our existentiell comportment in the world is always already fallen or inauthentic? It would seem that a way out of inauthentic fallenness cannot be an existentiell possibility—unless the authentic option, conceived on the existential level, is also existentiell.[4] This confusion concerning levels of argumentation becomes most apparent when one considers Heidegger's understanding of death. Suddenly, the ontological, which for the most part deals with the conditions of possibility for the ontical, is itself construed as an "ontical" possibility and vice versa. Hence the problem with Heidegger's analysis as Ricoeur sees it: The (ontical) resolve to face death (the ontological limit of all human possibilities) in the Stoic sense, when conceived as the only way to overcome our ordinary condition of inauthenticity, becomes an ontological category itself. Stoic resolve, then, seems to be the only existentiell possibility in the face of death because it is now thought to be the only way to achieve authenticity. But then, according to Ricoeur, Heidegger has not maintained the integrity of his own distinction between the theoretical (ontological) and the practical (ontical). If, in the end, the practical overturns the priority of the theoretical, why insist upon a distinction in the first place?

When Ricoeur says that "it is in this interval between the existential and the existentiell that a meditation on eternity and on death can be conducted," he is placing Heidegger's distinction within a Kantian context. For Kant, what stands between mind and world (existential and existentiell), as their mediation, is the productive imagination that schematizes the categories. Part of the reason why Heidegger confuses the existential and the existentiell, according to Ricoeur, is because he collapses their "between" too quickly. In Heidegger's philosophy, understanding (*Verstehen*) replaces Kant's imagination as that which mediates self and world. It does so by projecting Dasein into future, worldly possibilities for Dasein's being. The authenticity of this projecting is guaranteed when Dasein attempts to project itself against its "ownmost" possibility, which is death. But death is not so much an "imagined" possibility as it is an undetermined historical fact. Death awaits everyone; we know that as a matter of historical fact. So Heidegger's understanding, which replaces Kant's imagination, is given a factical orientation (i.e., the existential is confused with the existentiell).

But Ricoeur claims that imagination is more than this. What happens to the fictional quality of imagination in Heidegger's desire to link understanding and factical death? He collapses the middle (imagination) toward the historical/cosmological side of things too quickly. That means he overlooks the fictional/ phenomenological side of our thinking. Ricoeur's narrative understanding attempts to ameliorate Heidegger's neglect of the fictional component of imagination. As Ricoeur sees it, fiction is able to extend the domain of phenomenological inquiry through its imaginative variations. Fiction's ability to extend our thinking depicts its Kantian quality: For Kant, both the ideas of reason and religious symbols are able to extend philosophy beyond the strict boundaries constitutive for knowledge.[5] Ricoeur is suggesting that death, as Heidegger understands it, is a theoretical limitation (in the Kantian sense) for our thinking about time. But fiction is able to extend phenomenology/thinking beyond the strict confines of death, such that time's "other," eternity, can appear. By expanding Heidegger's understanding to include a fictional component, Ricoeur is able to retrieve a philosophical appreciation for Augustine's eternity.

There are two places in the extensive passage quoted above

where Ricoeur alludes to religious concerns. In the first place, he says

> For Augustine, this upper limit [to the process of temporalization] is eternity. And for the current in Christian tradition that incorporated the teachings of Neoplatonism, time's approximation of eternity lies in the stability of a soul at rest. . . . On the properly theological level [however], it is not certain that the conception of eternity is summed up in the idea of rest. We will not discuss here the Christian alternatives to the equating of eternity with rest.

What is especially interesting here is that Ricoeur is beginning to hint at a "deconstructive" reading of Augustine. That is, Ricoeur is attempting to put some distance between Augustine's Christian sense of eternity and the Neoplatonic sense of eternity. In the "Conclusion" at the end of Part IV, as we shall see in my last chapter, this deconstructive reading of the *Confessions* is made more explicit. There, Ricoeur argues that Augustine's sense of eternity is more biblical than philosophical.

In the second place, Ricoeur says

> For my part, I have unhesitantly held mortality to be a universal feature of the human condition. . . . But I left hanging the question whether the existential component of Being-towards-death, and perhaps even that of anticipatory resoluteness, leaves room for existentiell modalities other than the Stoic tone given by Heidegger to resolution, including the modalities of Christian hope stemming in one way or another from faith in the Resurrection.

Ricoeur wants to preserve the possibility of a resurrection faith, despite the ontological significance of death, by claiming that death need not be the only factor shaping our existential comportment. But he only attempts to preserve the possibility of faith on a philosophical level, not a religious one. As we shall see in chapter 4, Scharlemann argues that Ricoeur's hermeneutical theory cannot adequately interpret resurrection texts because it lacks a notion of textuality. He claims that Ricoeur's hermeneutics is still tied to Heideggerian existentialism, and neither Ricoeur nor Heidegger can account for a world where death is something in the past. So, despite a desire to preserve resurrection faith against ontological death, Ricoeur's neglect of religious issues keeps him from accomplishing it.

Western Classics on Time: Ricoeur's
Narratival Interpretation

Ricoeur's argument hinges upon his narratival interpretation of the classic reflections on time. Initially, he contrasts Augustine (*Confessions*) and Aristotle (*Physics*). Their respective theories— phenomenological for Augustine, cosmological for Aristotle— constitute the fundamental tension or aporia that repeats itself in every subsequent attempt to understand time theoretically. In a second stage, Ricoeur extends his narratival interpretation to the contrasting reflections on time made by Husserl and Kant. Only then, in a third stage, is he able to incorporate Heidegger into the flow of his antinomous narrative. We need to review these early contrasts/stages in order to substantiate his linkage of the *Confessions* with *Being and Time,* despite their historical separation.

Augustine and Aristotle

In the *Physics,* which consists of eight books, Aristotle inquires into the nature of nature (*physis*).[6] He defines nature as the principle of motion and change that is inherent to things. In Book III, he tries to determine the precise meaning of motion; and in Book IV, he deals with things such as place and time that are related to motion. The discussion of time entails the last five chapters (10–14) of Book IV. Let me review Aristotle's arguments before turning to Ricoeur's contrast of Augustine and Aristotle on time. (See the previous chapter for an extensive analysis of Augustine's theory of time.)

Aristotle scholars have identified a short, core treatise within the five chapters that deal with time.[7] It runs from the middle of chapter 10 to the middle of chapter 11 (218b9–219b2) and contains all the essentials of Aristotle's theory of time. The remaining parts of the text are thought to be appendices that are only loosely connected to the central argument and that reply to questions discussed in Aristotle's school. Ricoeur's reading of the *Physics* focuses on the core treatise.

It is interesting to observe that Aristotle begins his reflections on time the same way that Augustine does. He asks whether or not time belongs to the class of things that exist (ch. 10, 297). The same difficulties—with a past that no longer is, a future that has yet to be, and a present "now" that constantly passes away—preoccupy Aristotle. After rehearsing the difficulties, he suggests that time is an obscure phenomenon, the existence of which is difficult to pin down. His point is that the question of time's existence will be resolved only when the question of its nature is. Then comes the core treatise, which attempts to say precisely what time is. It develops the traditional view that time has something to do with motion and change.

First, Aristotle demonstrates that time is not identical to movement (or change). His reasoning is this: When something moves or changes, the change is peculiar to the thing itself. "But time," he says, "is present equally everywhere and with all things" (ch. 10, 298, 218b13). Time is universal, not peculiar to specific things. Clearly, then, it is not identical to movement. But Aristotle does not stop there. Even though time is not exactly movement, it cannot be independent of it either. For without a sense of change in things, we would have no sense of time. We never perceive movement without time passing or the passage of time without movement. So in order to discover what time is, we must discover what its relation to movement is. Then comes the critical phase of the argument. Aristotle defines time as the "number of motion in respect of 'before' and 'after' " (ch. 11, 299, 219b2). He suggests that our *awareness* of time depends upon our ability to distinguish a before and an after relative to something existent. The distinction between before and after implies a change in place and magnitude, otherwise we could not recognize a difference in things. This change or difference is brought about by movement; something has to move from A to B in order to establish a before and an after. Time is the marking or numbering of the movement that establishes the before and after in things.

> But we apprehend time only when we have marked motion, marking it by "before" and "after"; and it is only when we have perceived "before" and "after" in motion that we say that time has elapsed. Now we mark them by judging that A and B are different, and that some third thing is intermediate

to them. When we think of the extremes as different from the middle and the mind pronounces that the "nows" are two, one before and one after, it is then that we say that there is time, and this that we say is time. For what is bounded by the "now" is thought to be time—we may assume this. (Ch. 11, 299, 219a21–29)

Aristotle seems to be saying the following: Now I see this thing in this place. If I notice this thing in a different place, I know that time has elapsed and that there is a new "now" that is later. We have to "number" or recognize the movement in things in order to experience the passage of time. "Hence time is not movement, but only movement in so far as it admits of enumeration" (ch. 11, 299). Motion goes with magnitude, or the change in something as it changes place, and time goes with motion. Aristotle's theory depends upon this tripartite analogy. Note too how it depends upon human awareness.

In chapter 12, regarded as an appendix to the core treatise, Aristotle investigates what it means "to be in time." He compares it to being in number. " 'To be in number,' " he states, "means that there is a number of the thing, and that its being is measured by the number in which it is" (ch. 12, 301). Likewise, if a thing is "in time," it will be subject to measurement by time. That means it is affected by time and contained by it. Things grow old and decay because they are subject to time, because their being is in time. In chapter 13, Aristotle investigates the nature of the "now" and determines that it has a dual nature. The now both divides and unites things as the number of their motion or changes. The now is the boundary of before and after, past and future. In chapter 14, Aristotle considers how time can be related to the soul. As though giving prior consent to Augustine's later query, Aristotle says,

> Whether if soul did not exist time would exist or not, is a question that may fairly be asked; for if there cannot be some one to count there cannot be anything that can be counted, so that evidently there cannot be number; for number is either what has been, or what can be, counted. But if nothing but soul, or in soul reason, is qualified to count, there would not be time unless there were soul, but only that of which time is an attribute, i.e., if *movement* can exist without soul, and the

before and after are attributes of movement, and time is these *qua* numerable. (Ch. 14, 303, 223a21–28)

Since movement exists without the soul, time does as well *potentially.* But if time is the actual numbering of motion, then it requires the soul to do the numbering. So Aristotle, it seems, might be somewhat sympathetic toward Augustine's efforts to define time in terms of the soul.

As Ricoeur claims at the beginning of his contrast between Augustine and Aristotle on time, "Augustine did not refute Aristotle's basic theory of the primacy of movement over time, although he did contribute a lasting solution to the problem Aristotle left in abeyance concerning the relation between the soul and time" (III, 12). According to the cosmological tradition that Aristotle represents, time envelops us and dominates us in ways that exceed the powers of the soul. "I am convinced," says Ricoeur, "that the dialectic of *intentio* and *distentio animi* is powerless to produce this imperious character of time and that, paradoxically, it helps conceal it" (III, 12). Augustine's failure, according to Ricoeur, is that he tried to derive the principle of the extension and measurement of time from the distension of the mind alone. He could not substitute a psychological conception of time for a cosmological one. Both conceptions are necessary for a full understanding of time. The odd thing is that neither conception can allow for the other; they are involved in an unresolvable disagreement.

Ricoeur thinks that Augustine's refutation of the cosmological thesis is misdirected and simplistic. Recall that Augustine disagreed with the common view that time is nothing but the movement of heavenly bodies (ch. 23, 271). But such a claim cannot refute Aristotle since his thesis is that time has something to do with movement while not being movement itself. Thus Ricoeur claims that Augustine is forced to turn to the soul for his principle of extension far too quickly. Augustine oversimplifies the cosmological thesis, and his wager does not really pan out. What is the standard we use to compare the duration of the expectations and memories with which we measure time? Augustine cannot really say. And yet this notion of a fixed unit for comparing lengths of time is the basis for the cosmological theories that link time with motion as well. A "day" is the length of time it takes for the earth to make a complete rotation on its axis. The standard for measuring time is provided by the movement of the sun and the earth. The standard need not be identified with the sun's (or earth's) movement, it must be provided by it.

Ricoeur's claim is that Augustine's turn to the soul does not supercede the need to find a standard for measurement. In fact, Augustine exacerbates this problem because it is never clear what sort of access we can have to the impressions that are supposed to remain in the mind and indicate lapses of time. The problem of a fixed measure, which is central to the issue of measuring time, is not resolved by Augustine's inward turn. Hence his failure, and an opening for Ricoeur's comparative thesis:

> Augustine's failure to derive the principle for the measurement of time from the distension of the mind alone invites us to approach the problem of time from the other side, from that of nature, the universe, the world. . . . We shall later show how important it is for a theory of narrative that both approaches to the problem of time remain open, by way of the mind as well as by way of the world. The aporia of temporality, to which the narrative operation replies in a variety of ways, lies precisely in the difficulty in holding on to both ends of this chain, the time of the soul and that of the world. (III, 14)

Having demonstrated the faults in Augustine's theory, Ricoeur turns to Aristotle. He follows the convention I mentioned above, whereby the five chapters in Book IV of *Physics* are divided into a core treatise (218b9–219b2) and several appendices. Naturally, Ricoeur focuses on the core treatise, dividing it into a three-stage argument. The first stage extends to the point where Aristotle confirms that time is neither movement nor independent of movement (219a2). It is not identical to movement since time is everywhere while movement is particular to things that change. Nor is it unrelated to movement since we do not perceive the passage of time without also noticing the movement of something. Understandably, Ricoeur wants to minimize the role of the mind in this part of Aristotle's argument.

> This argument does not place particular stress on the mind's activity of perception and discrimination, or, more gener-

ally, on the subjective conditions of time-consciousness. The term that is stressed is "movement." If there is no perception of time without the perception of movement, there is no possible existence of time itself without that of movement. (III, 15)

Ricoeur thinks that the dependence of time on movement or change is an irrefutable, primitive fact. If we jump ahead, we could say that the task of narrated time, which is supposed to mediate between cosmological time and phenomenological time, becomes apparent precisely at this point. Somehow Augustine's distension of the soul must be grafted onto Aristotle's view that time belongs to movement. This difficulty goes to the heart of the problems concerning time.

The second stage in Aristotle's argument concerns the application of a before and an after relation to time. The relation between before and after is in time because it is first of all in movement, and time and movement go together. Further, the relation is in movement because it is first of all in magnitude, and movement and magnitude go together. So the before and after of a temporal relation depends upon the transfer of some existent thing (magnitude) from this place to that. Time has something to do with movement, specifically with the before and after inherent to movement. Ricoeur sees another irreducible, primitive fact in this part of Aristotle's definition. Aristotle's point, Ricoeur claims, is that

> succession, which is nothing other than the before and after in time, is not an absolutely primary relation. It proceeds by analogy from an ordering relation that is in the world before being in the soul. Once again we here come up against something irreducible. Whatever the mind contributes to the grasping of before and after—and we might add, whatever the mind constructs on this basis through its narrative activity—it finds succession in things before taking it up again in itself. The mind begins by submitting to succession and even suffering it, before constructing it. (III, 16) [8]

In the third stage, Aristotle completes his definition by adding a numerical relation to the relation between before and after. Time, he says, "is the number of motion in respect of 'before' and 'after' " (219bz). As Ricoeur notes, the argument once again rests on a feature of the perception of time. The mind must be able to distinguish the end points (the befores and afters) of a series of countable intervals. "But the privilege accorded movement is not weakened in any way by this" (III, 16). Ricoeur insists that Aristotle's definition of time contains no explicit reference to the soul, despite the fact that it draws upon operations of the soul such as perception. He does acknowledge an *implicit* phenomenology of time-consciousness in Aristotle's definition, however. (How can he avoid making that concession, since chapter 14 of the *Physics* deals explicitly with the soul?) "In truth, in one of the subsidiary treatises appended to his definition of time [chapter 14], Aristotle is the first to grant that the question of deciding whether 'if the soul did not exist time would exist or not is a question that may fairly be asked' " (III, 16).

The issue, for Ricoeur, is how Aristotle can refuse to include any phenomenological determination in the core definition of time. The answer is given in the definition itself: The soul's counting with respect to time is subordinate to *physis* and its principle of motion. I would make one observation about Ricoeur's reading of Aristotle to this point. It seems that once again Ricoeur's argument partially depends upon his division of a text. By claiming that the core treatise takes precedence over the "appendices," Ricoeur can make Aristotle's discussion of time fit his own argument, just as his earlier division of Augustine's Book XI of the *Confessions* enabled him to fit Augustine's discussion into it. Ricoeur is not without scholarly support in either of his divisions.[9] But perhaps too much rests on his practice of bracketing parts of texts in order to deny any sense of cosmology in Augustine and, now, any explicit sense of phenomenology in Aristotle.

According to Ricoeur's reading, the appended treatises actually represent Aristotle's failed attempts to overcome, in some sense, the overwhelming power of time. The cosmological thesis asserts the priority of time over the soul—we grow old and die because of the passage of time—and Aristotle cannot muster enough evidence to overturn such "immemorial wisdom." So there are attempts on Aristotle's part to develop a phenomenology of time, that is, to grant to the soul some power or control beyond the confines of physical time. But these are failed attempts. Just as, according to Ricoeur's thesis, cosmology cannot be derived from phenomenology (Augustine could not refute or displace Aristotle), we now see that the reverse is equally true: Phenomenology cannot be derived from cosmology (Aristotle cannot refute or displace Augustine either).

Ricoeur concludes his interpretation of the *Physics* by noting two persistent aporias in Aristotle's argument that tend to occlude the phenomenological aspect of time from his view. The first has to do with how difficult it is to conceive time because of its ambiguous nature. Time is caught between movement, of which it is an aspect, and the soul that discerns it. So which end of the stick does one pick up? By focusing on the priority of movement, Aristotle was forced to slight the soul. His attempts to mitigate that oversight failed. The second aporia is related to the first. It has to do with how difficult it is to conceive movement itself. It is no less ambiguous than is time. Ricoeur explains:

> Evoking the aporias proper to Aristotle is intended to show that he does not hold fast against Augustine owing to the strength of his arguments alone, but rather as a result of the force of the aporias undercutting his own arguments. For, over and above the anchoring of time in movement established by his arguments, the aporias these arguments run into indicate something about the anchoring of movement itself in *physis,* whose mode of being escapes the argumentative mastery that is so magnificently displayed in Book IV of the *Physics.* (III, 18)

The notion of *physis,* like that of time (or freedom, soul, etc.), can never be completely "caught" in any given conceptual framework. The very effort to do so engenders aporias, more and more difficulties.

A cosmological interpretation of time cannot adequately account for the soul, just as a psychological interpretation cannot adequately account for movement and physical change. Such is Ricoeur's conclusion after reading Augustine and Aristotle together. He pinpoints the difference between the two views of time in the contrast between an instant (Aristotle) and the present (Augustine). There is, he claims, a "conceptually unbridgeable gap between the notion of the 'instant' in Aristotle's sense [of time] and that of the 'present' as it is understood by Augustine" (III, 19).

> To be thinkable, the Aristotelian "instant" only requires that
> the mind make a break in the continuity of movement,
> insofar as the latter is countable. The Augustinian present,
> however, . . . is any instant designated by a speaker as the
> "now" of his utterance. It does not matter which instant is
> chosen, the present is as singular and as determined as the
> utterance that contains it. (III, 19)

The mind, necessarily, is involved in both conceptions of time.
But in the cosmological conception, the role of the mind (or soul)
is of secondary importance. It only breaks up the successive flow
of physical time in order to make it countable. It plays no
constitutive role in the phenomenon of time. Yet in the psychological (phenomenological) conception, time is not constituted by the
before-and-after succession of events. Instead, past and future are
designated as the before and after of a present self-reference.
Every utterance refers to the self and establishes a time from
which other times can be distinguished in terms of past and future.
How does one reconcile the cosmological instant and the lived
present? That is Ricoeur's question. The problems of time can
only be resolved by holding together the two different sides. Only
a poetics of narrative can join what speculation separates, according to Ricoeur. "Our narrative poetics needs the complicity as well
as the contrast between internal time-consciousness and objective
succession, making all the more urgent the search for narrative
mediations between the discordant concordance of phenomenological time and the simple succession of physical time" (III, 22).

Husserl and Kant

Ricoeur continues his investigation of the confrontation between Augustine's soul time and Aristotle's physical time by
turning to other classic texts. Confronting Husserl's *Phenomenology of Internal Time-Consciousness* with Kant's *Critique of Pure
Reason,* Ricoeur pushes the aporetics of time to a new level of
development. Husserl attempts to make time as such appear
within consciousness. He wants to free the phenomenological
investigations begun by Augustine from all aporias. But, according to Ricoeur, that ambition runs up against Kant's thesis that
time is essentially invisible. Husserlian phenomenology and
Kantian transcendentalism participate in the aporetics of time

initially articulated by Augustine and Aristotle. The latter dialectic opposed a time of the soul and a time of motion. In Husserl and Kant, that ancient opposition is transposed into a modern dialectic entailing interaction between subjectivity and objectivity.

The Phenomenology of Internal Time-Consciousness is based primarily upon lectures Husserl delivered in 1904–1905.[10] At that point, between publication of *Logical Investigations* (1901) and *Ideas Pertaining to a Pure Phenomenology* (1913), Husserl was continuing to develop his understanding of intentionality, a concept he had introduced, following Brentano, in *Logical Investigations.* In *The Phenomenology of Internal Time-Consciousness,* Husserl elucidates both the intentional character of time-consciousness and the temporal implications of any process of intending. "Intentionality" refers to the objective orientation of consciousness and mental activities in general. Consciousness is always consciousness *of* something; it always has an objective correlate. That means consciousness is self-transcending—it points toward a "transcendent" (nonmental) world.

Husserl wants to study the immanent or internal time (*innere Zeit*) that we experience as the flow of consciousness as such. Thus, his phenomenological analysis of time-consciousness brackets or disregards the objective time of nature, including its psychological effects. He carefully distinguishes the phenomenological apperception of time from the psychological apperception of time. The former stems from analysis of the lived experiences of time in consciousness itself, while the latter stems from analysis of how external stimuli produce temporal sensations in consciousness. In fact, Husserl develops his phenomenological theory of time-consciousness from criticisms he makes of Brentano's earlier psychological theory.

Brentano's theory attempts to explain how a melody is constituted by a succession of sounds.[11] He wonders whether we perceive such successions or whether we produce them. At issue is how we could ascertain a melody without being simultaneously aware of each note and of the effect of all the notes taken together. If each note simply persisted in consciousness as it sounded, there would be no memory of the whole. One note would displace the next, or worse, all the notes would sound together in a cacophonous mélange. According to Brentano, each tone must persist in a continuously modified state in order for all the tones to blend into

melodious harmony. Through a process of "primordial association," we produce fantasies of each sound that taper off with time
(eventually disappearing) so that the whole can be heard as a
melody of sounds. Brentano concludes that it is not possible to
perceive temporal phenomena such as succession, duration, and
alteration. Instead, we produce them as nonreal fantasies.

Husserl claims that Brentano's law of primordial association
has a psychological orientation since it pertains to the origin of a
mental sense of time in response to external stimuli. But he also
thinks there is a phenomenological core to Brentano's observa-
tions, which he intends to develop. The problem with Brentano's
theory, according to Husserl, is that the constitution of time is
relegated to the realm of fantasy. Brentano dismissed the possibility that we can have a perception of time in its primordial
associations; that is, he claimed our intuitions of sounds (or
objects) are extended into the past and future, thereby constituting
a unitary consciousness, by fantasy-induced associations. However, Brentano's perception versus fantasy question is too simplistic. It presupposes a distinction between conscious acts (primordial associations) and their content (the actual sounds one hears).
But it overlooks a third sphere in the constitution of time-
consciousness: the mental object apprehended in the conscious act
(i.e., the noematic correlate of the noetic activity). When this third
sphere is considered, it is possible to see the primordial associations that constitute time-consciousness as real experiences that
we can apprehend through phenomenological investigation. "In
fact," claims Husserl, "the whole sphere of primordial associations is a present and real lived experience" (*Phenomenology of
Internal Time-Consciousness,* 39). His theory attempts to explain
how that is the case.

Both Brentano and Husserl (like Augustine before them) focus
on phenomena of sound—how is it possible to hear the individual
tones of a melody as a unified whole?—in explaining time-
consciousness. The reason is simple. Hearing is like time-
consciousness because in both cases duration is an important
factor. Sounds endure for only a short time, then they pass away;
they flow like time itself. Motion is part of the phenomenon in
both cases. According to Husserl, when we hear a melody, we
actually perceive a "temporal object" (*Zeitobjekt*). Temporal
objects are unities in time that include temporal extension in

themselves. They appear to consciousness, or our lived experience immanently (phenomenologically) conceived, in a continuously modified form that slides or runs off into the past. So a melody, as a perceptible temporal object, is what Husserl calls a "running off [*Ablauf*] phenomenon." Each series of notes that we hear as individual impressions given in a "now" runs off into the past as a new series of notes comes into the now time of perceptive consciousness. This continuous modification or running off into the past creates the phased continuum that we are able to hear as a melody. According to Husserl,

> The sound begins and steadily continues. The tonal now is changed into one that has been. Constantly flowing, the *impressional* consciousness passes over into an ever fresh *retentional* consciousness. (51)

Temporal objects (such as a melody) include impressions (the givenness of present sounds) and retentions (the givenness of past sounds).

Retention, or primary memory, provides the run-off modifications that give time-consciousness its continuity. The perception of temporal objects includes more than just the now. "Truly," says Husserl,

> it pertains to the essence of the intuition of time that in every point of its duration (which, reflectively, we are able to make into an object) it is consciousness of *what has just been* and not mere consciousness of the now-point of the objective thing appearing as having duration. (53–54)

Retention, consciousness of what has just been, is the intentionality of time-consciousness. This is where Husserl improves upon Brentano's theory of primordial associations. Brentano never considered the role that retentional memory plays in the very constitution of temporal objects. We can perceive both the impressions that represent the present in its givenness and the retentions that represent the past in its givenness. Both are aspects of time-consciousness in its observable objectivity. Brentano turned to fantasy too quickly in his attempt to describe time-consciousness. The "primordial associations" that enable the individual notes to run off into a coherent whole are part of what is given (hence observable) in the phenomenon of hearing. Retention is not fantasy; as the run-off of perception (its "comet's tail"), it is part of the intentionality of perception. Brentano's fantasy is more closely associated with what Husserl calls "secondary remembrance" or "recollection."

Retention is joined to actual perception as part of its intentionality. Recollection is not directly linked to perception; it is the representation ("presentification") of an earlier perception-retention process. Recollection is memory in the traditional sense: We re-create or recall something that we experienced long ago. Retention, on the other hand, is memory in its givenness (rather than its productiveness) as part of our experience of the "just-now-past."

Temporal objects spread their content over an interval of time; they form a single continuum that is constantly modified. That means we cannot perceive them only in their present phase. We must take into account their passing into the past as well. Retention or primary memory enables us to do that. But there is a futural part of the "spread" (or "halo") of temporal objects as well. Husserl calls it "protention," and it constitutes a second feature of time-consciousness. Protention is different from retention, however, and less significant in Husserl's theory. It is an expectational aspect of secondary remembrance.

In the first section of *The Phenomenology of Internal Time-Consciousness,* Husserl critiques Brentano's theory of time-consciousness. In the second section, he develops a more sophisticated, nonpsychological understanding of Brentano's law of primordial associations. Finally, in the last section, he investigates how consciousness of temporal objects is constitutive of the flux of consciousness itself. Note that Husserl's investigations follow a pattern similar to Augustine's. In Augustine's case, thinking about time eventually led to a theory of the soul. Augustine discovered that the soul (*distentio animi*) was essentially a temporal phenomenon. In Husserl's case, we now see how thinking about time leads to a theory of consciousness. He discovers that consciousness itself, in its unitary wholeness, is constituted with time-consciousness. For Husserl, only when consciousness is engaged in the practice of following the flow of temporal objects can it be constituted in and for itself.

Husserl argues that a second intentionality inherent to the retention process enables the flux of consciousness to appear.

> Every shading off of consciousness which is of the "reten-
> tional" kind has a double intentionality: one is auxiliary to
> the constitution of the immanent object, of the sound. This is
> what we term "primary remembrance" of the sound just
> sensed, or more plainly just retention of the sound. The other
> is that which is constitutive of the unity of this primary
> remembrance in the flux [of consciousness]. . . . This is
> retention of the entire momentary continuity of continuously
> preceding phases of the flux. (106–7)

In the flow of time-consciousness, which follows the shadings-off
of temporal objects, a "pre-phenomenal" temporality manifests
itself. This is the flux of consciousness as such, which is indicated
as the second intention in the shading-off (retentional) process.
Thus, time-consciousness has two different aspects according to
Husserl. It intends an object, and it intends itself. Here, we see that
the basic aporia (subject/object) endemic to any reflection on time
appears in Husserl's theory.[12]

    Ricoeur's criticisms of Husserl's theory of time-consciousness
are primarily methodological. Phenomenological reduction attempts to bracket the outside world, yet it depends upon the
language and understanding that pertains to that world. There is a
paradox at the heart of the phenomenological project.

> The fact that the perception of duration never ceases to
> presuppose the duration of perception did not seem to
> trouble Husserl any more than did the general condition for
> all phenomenology, including that of perception; namely,
> that, without some prior familiarity with the objective world,
> the reduction of this world would itself lose its very basis.
> . . . The reverse side of this strategy . . . is the proliferation of
> homonymies, ambiguities in terminology, maintained by the
> persistence of the problematic of the perceived object under
> the erasure of intentionality *ad extra*. Whence the paradox of
> an enterprise based upon the very experience it subverts. (III,
> 25)

Husserl's analyses of inner time, according to Ricoeur, are largely
based upon a prior understanding of objective or world time.
Thus, the metaphors and phrases Husserl uses are homonymous
with those from ordinary language. "Would we use the expression
'sensed at *the same time*,' " asks Ricoeur, "if we knew nothing of

objective simultaneity, of temporal distance, if we knew nothing of the objective equality between intervals of time?" (III, 24)

Keeping the difficulties inherent to phenomenological method in mind, Ricoeur offers a detailed reading of *The Phenomenology of Internal Time-Consciousness*. He focuses on what he calls the two great discoveries of the Husserlian phenomenology of time: "the description of the phenomenon of retention and its symmetrical counterpart, protention, and the distinction between retention (primary remembrance) and recollection (or secondary remembrance)" (III, 25–26). The discovery of retention advanced our understanding of time by enabling us to see the present as more than just a now-point. Included in the present is the *duration* of a phenomenon as it falls off into the past. Thus, Husserl could solve the problem of the continuity of time without resorting to a synthetic (or imaginative) operation as did Brentano (and Kant). The distinction between primary remembrance and secondary remembrance advanced our understanding further. Memory has two different functions, depending upon how recent its concern in the past is. The "just passed" past is still part of the present as retention. The more distant past, however, is no longer an aspect of sensed temporal objects. The distant past must be recollected in imagination.[13]

The purpose of Husserl's *Phenomenology,* according to Ricoeur, is finally the same as every other phenomenology of time: to derive or constitute objective time from inner time. In Husserl's case, however, there is a rather ironic circularity. He brackets out the very thing he wants to constitute! Ricoeur claims that this methodological perversity tends to invert the priorities of relation. The phenomenology of internal time-consciousness hinges upon the recognition of something that endures. But such endurance derives more from the object than it does from consciousness.

Ricoeur turns to Kant in order to find "the reason for the repeated borrowings made by the phenomenology of internal time-consciousness with respect to the structure of objective time, which this phenomenology claims not only to bracket but actually to constitute" (III, 44). He demonstrates the way in which Kant's method refutes Husserl's claim that phenomenology is free of any reference to objective time and that it can disclose a temporality purified of any transcendent intention. But he also demonstrates how Kant's attempts to speak of a time that never appears as such borrows from an implicit phenomenology of time that is hidden by the transcendental mode of reflection. Let us turn to Kant's text, and then return to Ricoeur's interpretation.

Kant published the first version of his *Critique of Pure Reason* in 1781. In the "Transcendental Aesthetic," the first major section of the *Critique,* Kant investigates space and time as the principles of a priori sensibility. Sensibility is the mind's capacity to receive representations or intuitions of objects. The forms of any such representations (space and time) are found in the mind rather than in objects. Time is the pure form of inner intuition, space the pure form of outer intuition. In other words, time is the mode of our inner awareness, which is the subjective condition for all possible experience. Just how time functions in this regard is something Kant illustrates later, especially in the "Analytic of Principles" (which includes the "Schematism" and the "Analogies of Experience"). Kant claims that time, as the a priori condition of all appearances, has "empirical reality" and "transcendental ideality." To the extent that objects given to us are subject to intuition, time is real or objectively valid. All things as appearances, or as objects of sensible intuition, are in time. In that sense, time has empirical reality. But time has no absolute reality; it is not a property of things in themselves, but only of the mind in its sensible quality. In that sense, time has transcendental ideality.

In the "Schematism" and the "Analogies of Experience," Kant discusses the role of time in the constitution of knowledge. At issue in both sections is how intuitions are subsumed under the rule of categories or concepts through the process of judgment. The initial problem for Kant is to understand how judgments are able to combine such heterogeneous items as concepts and intuitions. So in the "Schematism" he asks,

> How, then, is the *subsumption* of intuitions under pure concepts, the *application* of a category to appearances, possible? A transcendental doctrine of judgment is necessary just because of this natural and important question. We must be able to show how pure concepts can be applicable to appearances. (A138/B177)

In order for subsumption (or judgment) to take place, the heterogeneity of concepts and intuitions must be overcome. For what is radically dissimilar cannot be combined or unified.

In the "Schematism," Kant demonstrates that there is a third thing that is homogeneous with (or similar to) both the category and the appearance. This third thing, which is both intellectual and (formally) sensible, he calls the "transcendental schema." The schema accomplishes a transcendental determination of time. As Kant says,

> Time, as the formal condition of the manifold of inner sense, and therefore of the connection of all representations, contains an a priori manifold in pure intuition. Now a transcendental determination of time is so far homogeneous with the category, which constitutes its unity, in that it is universal and rests upon an a priori rule. But, on the other hand, it is so far homogeneous with appearance, in that time is contained in every empirical representation of the manifold. Thus an application of the category to appearances becomes possible by means of the transcendental determination of time, which, as the schema of the concepts of understanding, mediates the subsumption of the appearances under the category. (A139/B178)

Kant is arguing that each category of understanding already contains formal conditions of inner sense or time that determine how the category will be applied to objects of intuition. These conditions are the schema of the concept (or category). The schemata are imaginative syntheses that outline a procedure or rule of understanding. They are not images but, rather, the procedural conditions the imagination must follow in providing images (intuitional representations) for concepts.

The schemata enable the employment of understanding by restricting its use. Part of that restriction includes the combination of consciousness into one unified apperception. Recall what I said earlier about Kant's discussion of time in the "Transcendental Aesthetic": Time is the condition of our inner state (as inner sense), which means that it is the mode of our awareness. Here, then, in the "Schematism," we begin to see just what Kant meant. The self (or "I think") that accompanies all our experience as its condition of possibility appears indirectly through the unity of apperception that is a product of schematism (or transcendental time determination). As Kant puts it,

> What the schematism of understanding effects by means of the transcendental synthesis of imagination is simply the unity of all the manifold of intuition in inner sense, and so indirectly the unity of apperception [or self-awareness] which as a function corresponds to the receptivity of inner sense. (B185)

After establishing that the pure concepts of understanding can be employed in a priori synthetic judgments through the help of a mediating schematism, Kant turns to the task of demonstrating how that employment is actually achieved. He argues that there are twelve transcendental principles for the use of understanding that correspond to the twelve categories. The principles of pure understanding direct the objective (or empirical) employment of the schemata or rules for the synthesis of concepts and intuitions. Thus, they are rules for the objective employment of the categories and the basis for the unity of apperception that makes experience (or knowledge) possible.

I focus only on the analogies of experience, which are the principles for the objective employment of the categories of relation.[14] There are three such analogies according to Kant: the principle of the permanence of substance, the principle of succession in time, and the principle of coexistence. Time as such, in its permanence or endurance that is the basis for its two possible modes (succession and coexistence), is at issue in the analogies. The general principle underlying all three analogies is that "experience is possible only through the representation of a necessary connection of perceptions" (B218).

Kant means by this that experience is an empirical knowledge, or a knowledge that determines an object through perceptions. The general principle of the analogies is that experience is only possible through representation of a *necessary* connection of perceptions. The principle rests upon the necessary unity of apperception, which is given in the transcendental determinations of time. So the analogies make experience possible in that they are rules pertaining to how appearances must be related in time. By means of these rules [the analogies] the existence of every appearance can be determined in respect of the unity of all time (B219). At issue is the time-order of the whole manifold of empirical consciousness, which must be united in the original apperception. Put succinctly, the analogies govern (or regulate) the employment of the schemata depicting the synthetic unity of all appearances. Since the unity of apperception is time-determined, the analogies pertain to the appearances of time in the relations of perceptions, even though time as such is invisible.

The first analogy is the principle of permanence in substance. Here, "substance" represents the permanence of time, which is not perceivable in itself. But we can perceive that substance is the permanent substratum of everything that exists. Thus, substance can indicate the backdrop for all succession and coexistence *in* time. As the backdrop for determinations of appearances in time, substance points to that which envelops all change—time itself as never changing. Kant's point is that change cannot be perceived unless it has a permanent background that cannot change. Otherwise, we would not be able to determine that change has occurred. The changes in the manifold of appearances that are expressed through time (in its modes as succession or coexistence) presuppose a time that will always remain. Only in the permanent (time itself) are relations in time (succession or coexistence) possible. According to Kant,

> Without the permanent, there is therefore no time-relation. Now time cannot be perceived in itself; the permanent in the appearances is therefore the substratum of all determination of time, and, as likewise follows, is also the condition of the possibility of all synthetic unity of perceptions, that is, of experience. (A183/B226)

In his reading of the *Critique,* Ricoeur shows that Kant is unable to construct the presuppositions for his invisible time without borrowing from an implicit phenomenology that is never expressed. Ricoeur claims,

> What most obviously opposes Kant to Husserl is the assertion of the indirect nature of all assertions about time. Time does not appear. It is a condition of appearing. (III, 45)

Ricoeur develops the contrast between Kant and Husserl by following the argument from the "Transcendental Aesthetic" to the "Schematism" to the "Analogies of Experience" (as I have done above). He focuses on the refutational style of Kant's arguments. Like Augustine, Kant employs the reductio ad absurdum. But unlike Augustine, Kant stands within the cosmic strand of reflections on time. Even though time is ascribed to inner sense (which would seem to indicate an affinity between Kant and Augustine), that sense is never the source for a self-knowledge. It is something we are only aware of phenomenally—that is, as it affects us in its objective employment.

According to Ricoeur, the unthematized phenomenological implication that runs through the *Critique* is found in references to the *Gemut,* or the affective sensibility, of the mind. But in the "Aesthetic," where *Gemut* is first discussed, Kant never appeals to its self-evidence. His arguments always proceed indirectly, by refutation of previous hypotheses. Thus, he always underscores the nonintuitive character of the properties of time. The emphasis is always on the presuppositional character of any assertion about time.

> This is why the discourse of the "Aesthetic" is that of presupposition and not that of lived experience. The regressive argument always wins out over direct vision. This regressive argument, in turn, assumes the privileged form of an argument from absurdity. (III, 47)

The paradox of the *Critique,* says Ricoeur, "is that its particular argumentative mode has to hide the phenomenology implicit in the thought-experiment that governs the demonstration of the ideality of space and time" (III, 48). Ricoeur turns to the "Analytic" in order to confirm that view, for that is where Kant demonstrates the necessity of the "detour by way of the constitution of the object" for any determination of time. The theme of the "Analytic" is that time, while not itself appearing, remains a condition for all objective appearing. The principles governing the imaginative operations of the schemata give only an indirect representation of time as they posit the conditions for the objectivity of objects. The indirect representation of time upon an object is also accompanied by a determination or lapse of time. But in both cases, the nonphenomenality of time is asserted over its intuitability. As the first analogy makes clear, we perceive objects *in* time, from which we can infer—but never "see"—that time as such is necessarily an enduring and permanent substratum. Since "time itself cannot be perceived, it is only by way of the relation between what persists and what changes, in the existence of a phenomenon, that we can discern this time that does not pass and in which everything passes" (III, 50). Ricoeur's conclusion about the *Critique* is that its implicit phenomenology cannot be articulated without breaking the reciprocal connection between the constitution of time and the constitution of the object. If that break were made, one would be led to Husserl's phenomenology of internal time-consciousness.

Ricoeur's confrontation of Husserl and Kant leads to the same impasse to which the earlier confrontation between Augustine and Aristotle led. Neither the phenomenological approach nor the cosmological (transcendental) one is sufficient by itself. Paradoxically, each borrows from the other while at the same time occluding the other. But the shift from Augustine and Aristotle to Husserl and Kant also marks a development (or at least a change) in thought. For Ricoeur,

> The polarity between phenomenology, in Husserl's sense, and critical philosophy, in Kant's, repeats, on the level of a problematic where the categories of subject and object—or more precisely of subjective and objective—predominate, the polarity between the time of the soul and the time of the world, on the level of a problematic introduced by the question of the being or nonbeing of time. (III, 57)

So there is a shift away from many of the metaphysical presuppositions inherent to the earlier conflict between Augustine and Aristotle. Both of them, for example, accepted the reality of the soul and the fact that it had some role to play in discovering the nature of time (a constitutive role for Augustine, a nonconstitutive role for Aristotle). Yet, there is no mention of a soul in Husserl's work, and such a notion is banished by Kant. Instead, there are now (with Husserl and Kant) references to inner states of mind, which are more or less available to observation. For anyone persuaded by the criticisms of metaphysics entailed in Kantian and post-Kantian philosophy, such changes represent a "development" in thought.

There is no question that Husserl stands in line with Augustine.

He acknowledges as much in the "Introduction" to his *Phenomenology*. Ricoeur has a more difficult time linking Kant with Aristotle. At first, one might suspect a closer relation between Kant and Augustine. Transcendental consciousness (specifically, the fact that time is understood as inner sense) seems to fulfill Augustine's philosophy of subjectivity. "But," says Ricoeur,

> this would be to forget the meaning of the transcendental in Kant, for its entire function lies in establishing the conditions of objectivity. The Kantian subject, we may say, is wholly taken up in making the object be there. . . . Time, despite its subjective character, is the time of a nature whose objectivity is wholly defined by the categorical apparatus of the mind. (III, 58)

Notes

1. To read Parts II and III as a supplemental detour is consonant with Ricoeur's stated intention to continue in *Time and Narrative* the discussion of the referential powers of language begun in *The Rule of Metaphor*. The issue of reference, or of a narrative text's ability truly to affect or refigure our practical experience, is of central importance. Hence, the precedence of Part IV in the argument.
2. See Rudolf Bultmann, "The Historicity of Man and Faith," in *Existence and Faith*, 92–110.
3. See "Heidegger's Argument" in the next chapter for a more detailed discussion of this problem concerning the complicity of Heidegger's two distinctions.
4. Heidegger terms it this way: "Factically, Dasein maintains itself proximally and for the most part in an inauthentic Being-towards-death. How is the ontological possibility of an *authentic* Being-towards-death to be characterized 'Objectively,' if, in the end, Dasein never comports itself authentically towards its end, or if, in accordance with its very meaning, this authentic Being must remain hidden from the Others? Is it not a fanciful undertaking, to project the existential possibility of so questionable an existentiell potentiality-for-Being?" (*Being and Time*, 304).
5. It goes without saying that Kant is of decisive importance in Ricoeur's thought. The notion that symbols give rise to thought, developed in *The Symbolism of Evil* and operative in *Time and Narrative* as well, is rather Kantian. As Ricoeur himself often points out in *Time and Narrative*, his extension of Aristotle's theory of emplotment is dependent upon Kant's insights into the imagination. The productive imagination of the "A" version of the "Transcendental Deduction" and the "Schematism" becomes, for Ricoeur, the operative in emplotment. The construal of a plot brings a manifold of episodic experiences under a rule of order much as a manifold of intuitions is imaginatively ordered when we understand a concept. In *Religion within the Limits of Reason Alone*, Kant speaks of religious symbols as imaginatively extending our thinking beyond

the strict limits of reason. See the discussion concerning the origin of evil on pages 34ff.

6. References to the *Physics* shall be to the edition found in Britannica's *Great Books of the Western World*, edited by R. Hutchins (*The Works of Aristotle: Volume I*, translated by R. P. Hardie and R. K. Gaye [259–355]. Chicago, IL: Encyclopaedia Britannica, Inc., 1952). Following quotations, I shall give chapter and page numbers, as well as international standard numbers where applicable.

7. Ricoeur makes this point in a footnote to his discussion of the *Physics* (*Time and Narrative*, III, 276, note 7).

8. At this point, Ricoeur is making an apology for Aristotle at Augustine's expense. I cannot help thinking, however, that Augustine would certainly agree with Aristotle—of course the mind begins by submitting to succession before constructing it. Why else would Augustine use three examples of hearing a sound in order to construct his psychological theory? The mind does not invent what it hears; it simply invents a way of understanding the order of what it hears. Such understanding depends upon one's having a good memory. Why have memory if the mind does not suffer what is beyond its creative control? If everything depended upon the mind, then imagination would suffice.

9. Ricoeur's reading of Aristotle's *Physics* depends heavily upon Paul Conen's *Die Zeittheorie des Aristotleles* (Munich: C. H. Beck, 1964). It is Conen's thesis that the treatise on time in the *Physics* has a core treatise to which are appended a series of less significant, smaller treatises. Conen's thesis is not without its critics, as Ricoeur notes. Victor Goldschmidt (*Temps physique et temps tragique chez Aristote* [Paris: Vrin, 1982]) is one critic to whom Ricoeur pays particular attention. Another point to make in this context is that Aristotle discusses the nature of time in places other than the *Physics,* such as in the *Poetics* and *Ethics.* So the cosmological definition does not exhaust Aristotle's thinking on the nature of time. Goldschmidt explores the similarities in Aristotle's different discussions of time, as Ricoeur mentions in a footnote (*Time and Narrative,* III, 280–81, note 27).

10. Actually, Husserl's *Phenomenology of Internal Time-Consciousness* has two parts, only the first of which pertains to the lectures of 1904–1905. But the second part is a series of supplements based upon later work by Husserl. So the first part constitutes a coherent text. It is the only part that Ricoeur interprets.

11. Recall how Augustine's theory of time also developed out of this context of interpreting the duration and succession of sounds. Both Brentano and Husserl expand upon Augustine's earlier work,

making more careful inquiries into the role that memory plays in establishing temporal experiences such as succession.

12. Husserl attempted to outrun the subject/object aporia by reducing the transcendent world to that which is immanent to the conscious mind itself. But the same difficulty reappears in Husserl's discussion of retention, which is the hallmark of his theory of time-consciousness.

13. The gap between these two kinds of memory is overcome when recollection is brought to bear on current lived experience; it becomes part of (intertwined with) the retentional consciousness that extends experience into the past.

14. The table of principles corresponds to the table of categories. The four groups of categories are: (I) Of Quantity; (II) Of Quality; (III) Of Relation; (IV) Of Modality. The corresponding principles are grouped thusly: (I) Axioms of Intuition; (II) Anticipations of Perception; (III) Analogies of Experience; (IV) Postulates of Empirical Thought in General. The three categories of relation are: (1) Of Inherence and Subsistence; (2) Of Causality and Dependence; and (3) Of Community. The three analogies of experience are: (1) Principle of Permanence of Substance; (2) Principle of Succession in Time (in accordance with the Law of Causality); and (3) Principle of Coexistence (in accordance with the Law of Reciprocity or Community). Even though Kant treats all the principles of pure understanding in their relation to inner sense (since their purpose is to establish the unity of apperception that is given in inner sense), I discuss only the "Analogies of Experience." I do so because that is where the three modes of time as such (duration, succession, and coexistence) are investigated.

Chapter 3
Heidegger's Being and Time

Ricoeur's thesis in Part IV of *Time and Narrative*—that the two traditions on time can be mediated only poetically, never theoretically—is the basis for his critical reading of Heidegger (just as it was for his readings of Augustine, Aristotle, Husserl, and Kant). He claims that Heidegger repeats the basic mistake of Augustine and Husserl, which is to occlude the cosmological significance of time. By placing Heidegger within the context of the two styles of thinking about time engendered by Western tradition, Ricoeur exposes weaknesses in the argument for *Sein zum Tode*. The weaknesses represent cracks in the buttress of ontological death, postcritical openings for Augustine's ancient claims about eternity stripped to their nonmetaphysical bone.

This chapter is divided into four sections. First, I offer an analysis of Heidegger's argument, focusing especially on his interpretations of death and time. Second, I make a case for my earlier assertion that Heidegger's notion of *Sein zum Tode* represents an inevitable outcome of the critical philosophy that began with Kant's *Critique of Pure Reason*. In making the connection between Kant and Heidegger, I rely upon Heidegger's *Kant and the Problem of Metaphysics*. In a third section, I review Ricoeur's assessment of *Being and Time*. I especially focus upon the implications that his critique of Heidegger holds for the postcritical revival of an Augustinian sense of eternity. Finally, in a fourth section, I briefly comment on Heidegger's enormous impact upon theologians.

Heidegger's Argument

Heidegger's task in *Being and Time*—to think the meaning of being—is conceived in terms of three steps or divisions. Otto Pöggeler has described the argument this way:

> In the first division of *Sein und Zeit,* Dasein is understood as
> Being-in-the-world and as care; in the second division, care
> is understood in its authenticity, and the sense of Dasein's
> Being as temporality. However, these two divisions serve
> only to prepare for the task which is set for the third division,
> to think the temporalness of the sense of Being.[1]

Only the first two divisions of *Being and Time* are published. The
text remains an unfinished fragment because Heidegger never
moved from his preparatory analysis of Dasein and temporality to
the goal of thinking about being. Nevertheless, the analytic of
Dasein stands on its own. The questions of death and time arise in
the second division of Heidegger's text.

Heidegger presents a "fundamental ontology" in *Being and
Time* that is inherently critical of traditional metaphysics. In order
to be "fundamental," ontology must be principally concerned with
the meaning of being.[2] That priority has been forgotten according
to Heidegger. Thus, his desire is to raise again the question
concerning the meaning of being that inspired Greek philosophy
(metaphysics), yet has been neglected since the time of Plato and
Aristotle. The unfolding of the argument in *Being and Time*
follows the criticisms that Heidegger makes of traditional metaphysics. Heidegger does not seek to annihilate or to displace
metaphysics. Rather, his criticisms are hermeneutical in the sense
that they are attempts to think what has been left unthought in
traditional metaphysics. What has been left unthought is the
fundamental question of philosophy—the question of the meaning of being. In order to rethink this question that originally
spawned metaphysics, claims Heidegger, one must question
behind the tradition of metaphysics.

Heidegger prepares to rethink the question of being by first
developing an existential analytic of Dasein. Dasein—
Heidegger's interpretation of the self—is the "there," that is, the
place of disclosure, for being. As the "there" of being in its
disclosure or advent, Dasein is privileged among entities (beings).
To have an understanding of being, even if it is an undeveloped
one, is a characteristic of Dasein's own being as the "there" of
being itself. Hence, in any concern that Dasein has for its own
existence, which is part of Dasein's daily, *ontical* fare, there is
(implicit) concern for the question of being. Ontical concern for

Dasein's own existence Heidegger terms "existentiell." Existentiell questions pertain to concrete possibilities of existing in this or
that manner, not to the ontological structure of existence. The
understanding that is concerned with the ontological rather than
the ontical Heidegger terms "existential." The existential analysis
of Dasein developed in *Being and Time* is an ontological en-
deavor.[3] Heidegger interrogates Dasein's existence in order to
exploit Dasein's preunderstanding of being, thus developing his
fundamental ontology: "Therefore fundamental ontology . . . must
be sought in the existential analytic of Dasein" (34 [13]).

Heidegger's insight is just this: To the extent that the problem
or question of being has been neglected, Dasein, in its unique
status as the "there" of being, has also been neglected. By
investigating the latter, the groundwork for the former can be laid.
The question of being, says Heidegger, is "nothing other than the
radicalization of an essential tendency-of-Being [*Seinstendenz*]
which belongs to Dasein itself—the pre-ontological understanding of being" (35 [15]). So Heidegger's investigation of Dasein,
coupled with an investigation of time, is (supposedly) able to raise
the question of the meaning of being. He argues that temporality
"links" the neglect of being and the neglect of Dasein since the
meaning of Dasein's being, as we shall see later, *is* temporality.
Since comportment toward being is a basic aspect of Dasein as
temporal being, time becomes the principal linking phenomenon
between the analytic of Dasein and the prospective discussion of
the meaning of being as such.

Dasein's nature is revealed in its unique status as the "there" of
being. To be Dasein is to be in "disclosure" (or disclosedness,
[*Erschlossenheit*]); disclosure is the characteristic mode of
Dasein's being. Dasein comes to itself through disclosures, or
through delimitations that "catch" Dasein in the act of existing.
Hence, in the preparatory analysis of *Being and Time* (division
one), Dasein's structure is described as being-in-the-world (*in-
der-Welt-sein*). For it is as a worldly entity, an entity always
already defined by its interactions within a world of concerns, that
Dasein finds itself being disclosed. "Dasein has being-in-the-
world as its essential state," states Heidegger (80 [54]). Without a
world, or a context, Dasein's self-disclosure would not be possible. Whenever something of Dasein is disclosed, what gets
revealed is Dasein's concernful relations to entities in a world.

Heidegger argues that knowing, which we commonly think of as
defining our relations with things, is a derivation of Dasein's more
basic state of being-in-the-world.

> Being-in-the-world, as concern, is *fascinated by* the world
> with which it is concerned. . . . [Thus] knowing is a mode of
> Dasein founded upon Being-in-the-world. Being-in-the-
> world, as a basic state, must be Interpreted *beforehand.* (88,
> 90 [61, 62])

When we "know" things, or just perceive them as objects, we
actually have to suspend their importance in the constitution of
our world. Objectivity is a derivative state for Dasein, which first
comes into disclosure as concerned for the world.

Heidegger discusses four principal disclosures of Dasein understood as being-in-the-world: mood or state of mind (*Befindlich-
keit*), understanding (*Verstehen*), discourse or language (*Rede*),
and fallenness (*Verfallenheit*). The "thrownness" or sheer facticity of Dasein's being-in-the-world is disclosed in Dasein's
fundamental moods. On the other hand, Dasein's free possibilities
for being-in-the-world (as opposed to its factical givenness) are
disclosed in projections of understanding. Heidegger often comments that every mood has its understanding, and every understanding its mood; these two disclosures constitute the equiprimordial kernel of Dasein's being-in-the-world. "The fundamental
existentialia which constitute the Being of the 'there,' the disclosedness of being-in-the-world, are states-of-mind and understanding" (203 [160]). Discourse is what links, or holds together,
the disclosures of mood and understanding. It is the medium of
articulation for both feeling and understanding; in that sense, it
too is an equiprimordial disclosure of Dasein.[4] Despite the
equiprimordiality, however, understanding is given priority as
Dasein's basic disclosure. Understanding, for Heidegger, is an
imaginative affair. It involves taking something for or "as"
something else through imaginative projections that play out
different possible options for Dasein's being.[5]

To be in disclosure or to be disclosure as such, as Dasein is,
means that Dasein is a "thrown project." Dasein is an entity whose
constitution is something already given ("thrown"), yet Dasein
must appropriate that constitution as its own through its efforts to

understand ("project"). Dasein's disclosure can become self-disclosure if Dasein follows the call that comes to it from its own being-in-the-world. The problem is that Dasein mostly refuses to listen to this call to self-understanding, or Dasein listens to the "wrong" call, getting lost in the idle chatter of the "they." For the most part, then, Dasein is fallen—mistaken in its self-understanding—because it refuses to understand itself as it truly is.[6]

Heidegger claims that in its fallen state, Dasein refuses to be what it is, a "thrown project." This refusal to become what it already is entails a dispersion of Dasein taken as a whole. To be fallen disrupts Dasein's integrity: How can Dasein integrate its basic moods and projections through self-communication if Dasein is constantly slipping out of its responsibilities? Lost in its distractions with the "gossipy they," fallen Dasein is inauthentic. ("Dasein's facticity is such that *as long as* it is what it is, Dasein remains in the throw, and is sucked into the turbulence of the 'they's' inauthenticity" [223 (179)].)

Two interrelated questions, consequential to Dasein's fallen-ness, focus the transition to division two of *Being and Time*. The first is: How is it possible to conceive Dasein as a whole or in its totality? The second: How can Dasein be conceived in its authenticity?[7] Heidegger argues that the unity or wholeness of Dasein must be understood (ontologically) as care (*Sorge*). That means care defines Dasein's kind of being as already thrown into a world in which self-projection (or self-understanding) is a necessary obligation. Dasein's tendency to be inauthentic, to succumb to its own fallenness, perverts care as Dasein's authentic call to be an integrated whole. Or, in its fallen state of disclosure, Dasein disrupts the care that constitutes its own being-in-the-world through integration of Dasein's moods, understanding, and discourse. In the second division of *Being and Time,* then, Heidegger pursues these two related questions that arise from the realization of Dasein's fallen nature—the question of Dasein's wholeness or unity understood as care and the question of Dasein's authentic, nonfallen state. How to conceive Dasein as a whole and how to call Dasein back from its fall into the inauthentic "they" are two aspects of the same problem. Fallen Dasein always refuses to be the whole that it is.

In the second division, Heidegger argues that Dasein is turned back from its lost venture into the "they" when it listens to the call of its own conscience. Conscience is the key to Dasein's *authentic* disclosure; it is disclosure as such, or the call of care itself as the integrating totality of Dasein's basic disclosures (mood, understanding, discourse). In Heidegger's words:

> Conscience [*Gewissen*] discloses, and thus belongs within the range of those existential phenomena which constitute the *Being of the "there"* as disclosedness. We have analyzed the most universal structures of state-of-mind, understanding, discourse and falling. If we now bring conscience into this phenomenal context, this is not a matter of applying these structures schematically to a special "case" of Dasein's disclosure. On the contrary, our Interpretation of conscience not only will carry further our earlier analysis of the disclosedness of the "there," but it will also grasp it more primordially with regard to Dasein's authentic being. . . . Conscience summons Dasein's Self from its lostness in the "they." (315, 319 [270, 274])

What the call (*Ruf*) of conscience reveals is a resoluteness (*Entschlossenheit*) on the part of Dasein. "Resoluteness signifies letting oneself be summoned out of one's lostness in the 'they' " (345 [299]). Dasein resolves to take the guilt of its own fallenness upon itself when it hears the call of conscience.

Furthermore (and parallel to the question of authenticity), Heidegger argues that Dasein can be grasped as a whole only when conceived against the backdrop of its uttermost possibility. Remember that Dasein, as "thrown project," is always already ahead of itself, intimately involved in a world with other entities. It is because Dasein's nature is to be ahead of itself, which means Dasein is always oriented toward what is "not yet," that its own being can be an issue for it. Dasein must choose itself in order to be itself; future possibilities—the conditions of possibility for any free choices—play a decisive constitutive role for Dasein's nature. The disclosure of Dasein's possibilities, always in light of the prior givenness of Dasein's mood, is a function of understanding and interpretation. The tendency to fall or slip away from authenticity results from Dasein always being ahead of itself and involved with its own possibilities. But the problem here is to grasp Dasein in its totality. How does one grasp what is always ahead of itself? The only way, according to Heidegger, is for Dasein to anticipate the *end* of all possibilities, to view itself from its own furthest limit.

Only when Dasein's involvement with what is ahead is fully anticipated, and brought fully into play in Dasein's existence, can Dasein be grasped as a whole. The mood for Dasein's anticipation (*Vorlaufen*) of its uttermost limit is anxiety (*Angst*); the end to be "understood," of course, is death. Only in *Sein zum Tode*, fully anticipating death as its "ownmost" possibility, can care, the totality of Dasein, be grasped as a whole. What fallen Dasein flees is precisely the thrownness of Dasein into anxiety and death as Dasein's ownmost possibility. How can Dasein overcome its tendency to flee? As I mentioned above, it must listen to its own call of conscience and resolve to live in full anticipation of its ownmost possibility: death. "Anticipation makes it manifest that this entity [Dasein] has been thrown into the indefiniteness of its 'limit-Situation' [death]; when resolved upon the latter, Dasein gains its authentic potentiality-for-Being-a-whole" (356 [308]). Heidegger is able to respond to the two principal consequences of Dasein's fallenness (disunity and inauthenticity) by combining the *anticipation* that answers the question of the unity or wholeness of Dasein with the *resoluteness* that answers the question of Dasein's authenticity. "Since resoluteness is constantly certain of death—in other words, since it anticipates it—resoluteness thus attains a certainty which is authentic and whole" (356 [308]).

"Anticipatory resoluteness" (*vorlaufende Schlossenheit*) is the complex notion with which Heidegger finishes his existential analytic of Dasein. "Dasein becomes 'essentially' Dasein in that authentic existence which constitutes itself as anticipatory resoluteness" (370 [323]). With the totality of Dasein's structural whole in hand, Heidegger is finally free to interpret the meaning of care as such (or Dasein in its totality). Meaning, for Heidegger, is what makes the understanding of something possible.[8] It underlies the process of projection through which understanding discloses itself. Every projection of possibility must be projected upon something; the "upon which" of any projection is the meaning underlying it. The meaning of care, according to Heidegger, is time. An explanation of the way in which he arrives at this conclusion is in order.

Dasein, as always already ahead of itself, is futural. Being-toward-death, anticipating death, makes Dasein's authenticity a futural phenomenon. In its nature as thrown entity (mood), however, Dasein has the character of "having been." Taking over the guilt and anxiety inherent to its thrownness, Dasein resolves to be as it already was. But the only way for Dasein to "take over" its own thrown nature is to project itself upon its uttermost possibility: death. The unity (or circle) of Dasein's being-in-the-world is temporally constituted when the future instantiates the present through the past. Heidegger describes it thus:

> As authentically futural [*eigentlich zukünftig*], Dasein *is* authentically as "having been" [*gewesen*]. Anticipation of one's uttermost and ownmost possibility is coming back understandingly to one's ownmost "been." Only so far as it is futural can Dasein *be* authentically as having been. The character of "having been" arises, in a certain way, from the future. . . . Coming back to itself futurally, resoluteness brings itself into the Situation by making present [*gegenwärtigend*]. The character of "having been" arises from the future, and in such a way that the future which "has been" (or better, which "is in the process of having been") releases from itself the Present. This phenomenon has the unity of a future which makes present in the process of having been; we designate it as "*temporality*" [*Zeitlichkeit*]. Only in so far as Dasein has the definite character of temporality, is the authentic potentiality-for-Being-a-whole [*Ganzseinkúnnen*] of anticipatory resoluteness, as we have described it, made possible for Dasein itself. *Temporality reveals itself as the meaning of authentic care. . . . The primordial unity of the structure of care lies in temporality.* (373–75 [326, 327])

Dasein's future, having been (or past), and present are "ecstases" of its own care understood as temporality. That means Dasein is an entity essentially "ecstatic" to itself or outside itself (recall that Dasein is a "thrown project"), which is a constitution it shares with time. Both time and Dasein are ecstatic phenomena, which is why time can be the meaning of care. In fact, "primordial time," for Heidegger, is temporality as such, that is, the process of temporalizing Dasein (whereby Dasein becomes what it already was) in the unity of Dasein's ecstases. So the meaning of Dasein, according to Heidegger, is temporality. Clearly, in that assessment—that time is of the essence of the self—Heidegger stands upon the shoulders of Augustine.

In the last three chapters of division two, Heidegger further develops his notion of primordial time as the temporality of Dasein. The first step in this furtherance of the argument is to repeat the analysis of Dasein presented in division one in order to reveal Dasein's average, inauthentic kind of being—everydayness—in its temporal meaning. The second step is to investigate the meaning of the historicality of Dasein. And the third is to develop a concept of "within-time-ness" (*Innerzeitigkeit*) in order to broach the ordinary, traditional conception of time as that within which things have their being. The purpose in all three chapters is to show that Heidegger's interpretation of temporality is the condition of possibility for our everyday experiences of time and for our traditional understandings of history and time.

From Kant to Heidegger

Death is the ownmost possibility that lies ahead for every Dasein. By seizing that possibility, Dasein is enabled to achieve authenticity. Included in Dasein's authentic mode is a unity of its temporal being-in-the-world. Dasein's temporality (*Zeitlichkeit*), in other words, is defined by its own finitude. Death is the limiting horizon against which the totality of care, or Dasein in its worldliness, can manifest itself. All this has been discussed. But some demonstration as to why the depiction of death in Heidegger's *Sein zum Tode* has an increased ontological significance must be made. The claim of this study is that *Sein zum Tode* best illustrates the "ontological death" that has recently threatened the very possibility of any religious (or transcendent) meaning. That claim can be justified by demonstrating how *Being and Time* fulfills the criticism of a transcendent, metaphysical realm that Kant's philosophy inaugurated.

The reading of the *Critique of Pure Reason* that Heidegger offers in *Kant and the Problem of Metaphysics* has been criticized for being self-serving and idiosyncratic.[9] But whether Heidegger's interpretation of Kant from the perspective of finitude and transcendental imagination can be justified is not the issue here. What is at issue is the fact that Heidegger conceives his own thought as a radicalization of Kant's criticism of traditional metaphysics. So Heidegger himself thinks that the fundamental ontology he develops in *Being and Time* is an extension of Kant's *Critique*. Heidegger's attempt to think the meaning of being within the constraints of human finitude is inspired by Kant's discussion of the transcendental imagination (in the "A" version).

Heidegger interprets the *Critique* "as a laying of the foundation of metaphysics in order thus to present the problem of metaphysics as the problem of a fundamental ontology."[10] Metaphysics is inquiry about being; as such, it seeks transcendent knowledge, or knowledge of the supersensible. How can such a project be justified? A foundation must be laid that ultimately depends upon the capacities and limitations of human thought. Fundamental ontology, as inquiry about Dasein (the existent, finite being for which the question of being is a natural predilection), is the basis for metaphysics. Thus, according to Heidegger's reading, Kant's question as to how a priori synthetic judgments are possible is his formulation of the problem of fundamental ontology. Kant argued that knowledge is constituted by a synthesis of intuitions from our faculty of sensibility and concepts from our faculty of understanding. Yet Heidegger insists that the two fundamental sources of knowledge (intuition and concept) actually arise from an original unity given in the faculty of transcendental imagination. Heidegger offers a reading of the *Critique* that attempts to uncover what Kant was groping toward: explication of the essential *finitude* of human reason and knowledge.

At issue in Kant's question about the possibility of a priori synthetic judgments, according to Heidegger, is the *finitude* of human reason and knowledge. Knowledge is an act of representation (something presenting something else). It can be either intuitive or conceptual, that is, relating immediately to a single object or relating mediately through some general feature. There is an evident symmetry: One could say that knowledge is a thinking intuition and that it is an intuitive thinking. Heidegger does not want to dispute the obvious reciprocity, yet he insists that, for Kant, intuition defines the true essence of knowledge. Human cognition is primarily intuition; in other words, cognition is dependent upon the reception of sensible intuitions. Only as essentially intuition can knowledge be seen in its finitude. But intuitions must be understandable, which means that they must be represented in a determined and general way through acts of judgment. Thinking (understanding) unites with intuition in order to serve it. The unity of sensibility and understanding, given in transcendental imagination, delimits the essential ground of finite knowledge in its receptive quality. Hence, according to Heidegger, the *Critique* must be read from the perspective of the "Schematism," or the synthesis of pure intuition (time) and the categories in imagination. The transcendental determinations of time characterize the receptivity or finitude of all knowledge. As Heidegger says,

> Kant's laying of the foundation of metaphysics leads to the transcendental imagination. This is the common root of both stems, sensibility and understanding. As such, it makes possible the original unity of [apperception]. This root itself, however, is implanted in primordial time. The primordial ground which is revealed in the Kantian laying of the foundation is time. (207)

According to Heidegger, Kant understands "primordial time" as the pure self-affection (or the very ability of the self to be affected, taken as an unconditioned whole) that constitutes the finite self as such. Time "gives itself"—it becomes what it always is—as the condition of the possibility for any sense of self or any inner affection. It is the self-giving of pure receptivity, which defines the transcendental ego as such. This primordial self/time makes the schemata or transcendental time determinations of the transcendental imagination possible. The self in its pregiven unity is temporal—such is the conclusion that Heidegger takes from the *Critique,* and further develops in *Being and Time.* Thus, Dasein is seen as a temporal unity of mood, understanding, and discourse. Finitude or death constitutes its ultimate horizon. Thus, the significance of death in Heidegger's philosophy is linked to Kant's critical philosophy. In order to lay the foundation for a chastened metaphysics, Dasein—the one whose very being is determined by the question of being—must be understood in its finitude.

The link between the *Critique of Pure Reason* and *Being and Time* indicates the utter significance of an ontological understanding of death for theological reflection. Heidegger's finitude

radicalizes Kant's criticisms of metaphysics, including metaphysical theology. Heidegger's understanding of ontological death
represents a threat to religious meaning that is similar, though
perhaps more radical and virulent, to Kant's understanding of
transcendental illusion. Whereas the latter defies a certain mode
of rational transcendence (to argue that God exists because God is
a necessary idea), the former seems to threaten the very possibility
of transcendence. For if Dasein is absolutely bounded by its own
finitude, then nothing can escape the confines of time.

Ricoeur's Assessment of Heidegger

Ricoeur's interpretation of *Being and Time* in *Time and Narrative* constitutes one chapter ("Temporality, Historicality, Within-
time-ness: Heidegger and the 'Ordinary' Concept of Time") in
Part IV. It follows the interpretation of Husserl and Kant that I
discussed at the end of the last chapter. Remember that Ricoeur
divided the classic discussions of time into two strands. One,
stemming from Aristotle and running through Kant and contemporary science, offers an objectivist perspective on time. The
other, stemming from Augustine and running through Husserl and
now Heidegger, offers a subjectivist perspective on time. According to Ricoeur, the aporias that are a permanent feature of any
philosophical reflection on time result from the incompatibility of
the two perspectives. In each case, whether phenomenological or
cosmological, the other perspective is occluded from consideration even while being tacitly presupposed. Both perspectives are
necessary in order to understand time, claims Ricoeur. A poetic
schematism, promoting a third, mediating perspective on time,
can overcome the gap between the two reflective positions.

Ricoeur's interpretation of *Being and Time* is conducted in
three parts that reflect the three "admirable discoveries" by means
of which Heidegger's hermeneutical phenomenology extends our
understanding of time beyond Augustine and Husserl. The first
discovery pertains to Heidegger's ability to envelop the question
of time in the care structure of Dasein. The second pertains to
Heidegger's construal of the three dimensions of time (future,
past, present) as an ecstatic unity. The third concerns Heidegger's
hierarchization of the temporalization process that stems from the

ecstatic unity of time. The three levels of the hierarchy are depicted in the concepts of temporality, historicality, and within-time-ness. Heidegger's basic mistake, according to Ricoeur, is that he attempts to understand time in terms of *Sein zum Tode.* The mistake is associated with the first discovery mentioned above, as I shall demonstrate in a moment. Ricoeur argues that this difficulty generated within the first discovery is taken up and multiplied by the second and third discoveries. Thus, his reading of Heidegger demonstrates the thesis that the more refined a phenomenological inquiry into the nature of time becomes, the more difficulties (aporias) it uncovers and generates. A pure phenomenology of time is just not possible.

In *Being and Time,* the question of temporality arises by way of the question of the possibility for Dasein's being a whole. And, as addressed above, the conception of Dasein as a whole is tantamount to understanding the care structure of Dasein's being-in-the-world. Thus, the structural wholeness of time itself—specifically, the unity of future, past, and present—is the question of first priority concerning time in Heidegger's hermeneutical phenomenology, and it is enveloped within the issue of Dasein's care structure. Given this intimacy of Dasein's care structure and its own temporality, it is appropriate to conceive the question of Dasein's wholeness in terms of an existential-temporal unity; both the existential and the temporal are implicated in the question of unity (care). Whereas Augustine articulated the unity of time's three modalities by enhancing the importance of the present (future things are made present in expectation, past things are made present in memory, present things are made present in attention), Heidegger does so by enhancing the importance of the future. The present, according to Heidegger, is too closely associated with fallen forms of existence to be the basis for an articulation of Dasein's existential-temporal unity. (Remember that what is ontically closest to us—the present, or things present-at-hand—is ontologically farthest away due to our tendency to be fallen.) The authentic unity of care is instead found in Dasein's (futural) tendency to be ahead of itself. But in order to grasp the "ahead of itself" in Dasein's unity, the limit or end of Dasein's being-ahead must be anticipated. Hence Heidegger's suggestion that *Sein zum Tode* is constitutive for Dasein's totality. Here is where the difficulties with Heidegger's text begin to arise for Ricoeur.

> This entrance into the problem of time through the question
> of Being-a-whole and this *alleged* connection between
> Being-a-whole and Being-towards-death pose an immediate
> difficulty, which will not be without effect on the other two
> phases of our analysis. This difficulty lies in the unavoidable
> interference, at the heart of the analytic of Dasein, between
> the existential and the "existentiell." (III, 64; italics added)

Ricoeur claims that the distinction between the existentiell and
the existential (or between the ethical and the theoretical) is
obscured by its interference with the distinction between the
authentic and the inauthentic. Heidegger's phenomenology, he
says, "is continually obliged to provide an existentiell attestation
for its existential concepts" (III, 65). The reason the testimony of
the existentiell is required in the existential analysis is because
part of the latter task is to overcome the inauthenticity characteristic of Dasein in its everydayness (or in its ordinary existentiell
comportment). Without a guarantee of authenticity, the existential
analysis of Dasein always falls short of primordiality. The relation
between being-a-whole and being-towards-death provides the
most striking example of this complicity concerning the two
distinctions. On the existential level, the only way that being-
towards-death can be assured of its constitutive role in the
existential-temporal unity of Dasein (understood in terms of
Dasein's being-a-whole) is if it can be coincidentally assured of
its own primordiality by means of the authentic attestation of
conscience on an existentiell level.

Ricoeur suggests that Heidegger's resoluteness in the face of
death constitutes the supreme test of authenticity only within a
Stoic ethical (or existentiell) configuration. Other ethical/
existentiell conceptions of authenticity are equally legitimate. In
other words, Ricoeur thinks that Heidegger's conception of death
as the uttermost possibility inherent to the structure of care is due
to a "recoil-effect" of the existentiell on the existential. Thus his
use of the word "alleged" in the above quotation. The peculiar
existentiell mark of Heidegger's "existential" analysis of temporality—that it is primordially determined by *Sein zum Tode*—has
serious consequences. The most serious consequence for Ricoeur
concerns Heidegger's discussion of the hierarchization of temporality in the last chapters of *Being and Time*.

> Despite the desire to derive historicality and within-time-ness from radical temporality, a new dispersion of the notion of time will, in fact, emerge from the incommensurability of mortal time, which temporality is identified with by the preparatory analysis, historical time, which historicality is supposed to ground, and cosmic time, which within-time-ness leads to. (III, 67)

Ricoeur's point here is blunt: Historicality and within-time-ness cannot be derived from a primordial temporality grounded in death.

Let us recall Ricoeur's argument concerning the three different conceptions of time. First, there is phenomenological time associated with human selfhood; second, there is cosmic time associated with the natural world; and, third, there is historical (or narrated) time associated with narrative identity. The historical time of narratives bridges the irreconcilable gap between phenomenological time and cosmic time. Each conception of time has its own primordial integrity. As the quotation in the above paragraph indicates, Ricoeur's three conceptions of time are actually derivable, with modification, from Heidegger's model of hierarchization. Heidegger's understanding of primordial temporality, minus its grounding in death, is a model for Ricoeur's own sense of phenomenological time. Phenomenological time is projective (futural) in Ricoeur's estimation, which of course follows from Heidegger. That is why Ricoeur can link fiction so closely to phenomenology. Ricoeur's "imaginative variations" on fictive plots, which present potentially inhabitable worlds to readers, constitute a complex application of Heidegger's conception of understanding as imaginative projection of Dasein's possibilities. Likewise, Heidegger's understanding of within-time-ness is a partial model for Ricoeur's sense of cosmic time, and Heidegger's historicality is a partial model for Ricoeur's narrative time. Again, according to Ricoeur, the Heideggerian model is mistaken in its attempt to derive historicality and within-time-ness from primordial temporality. Such derivation overlooks the incommensurability of the phenomenological (primordial temporality) and the cosmological (within-time-ness).

Heidegger's difficulties begin with his identification of primordial temporality and *Sein zum Tode*. To identify the "ahead of

itself" nature of Dasein's own temporality with death already
constitutes a tacit move to occlude the gap between phenomenological time and cosmic time. Death is something factical, historical. In an effort to make cosmic time (our ordinary understanding
of time) derivative, Heidegger takes something associated with
historical time (death) as constitutive for phenomenological time
(temporality). The reason this move on Heidegger's part is an
attempt to occlude the gap between the phenomenological and the
cosmic is because the historical is more "like" the cosmic than is
the phenomenological. History, like cosmic time, is able to
envelop us in that which is beyond our control. This "objective"
feature of time is introduced into the phenomenological understanding when death is associated with primordial temporality.
Thus, Heidegger's misunderstanding of the differences among the
three conceptions of time is a result of his misinterpretation of
primordial temporality in terms of death.

From Ricoeur's perspective, death is more associated with
historical time than with phenomenological time because an
existentiell attestation is necessary to insure its existential meaning. Existentiell attestation amounts to historical evidence in this
case. Heidegger's identification of the projective "ahead of itself"
nature of phenomenological time with a historically attested *Sein
zum Tode* is not able to insure the derivation of our ordinary
conception of time. The latter is grounded in something independent of any phenomenological conception, namely, cosmic time.

> One hypothesis was excluded from the outset by Heidegger:
> that the process held to be a phenomenon of the leveling off
> of temporality [or the derivation of ordinary time from
> phenomenological time] was also, and simultaneously, the
> separating out of an autonomous concept of time—cosmic
> time—that hermeneutic phenomenology never completely
> follows through on and with which it never manages to come
> to terms. (III, 88)

One can now understand just what Ricoeur means when he says
that he wants to think of death and eternity (Heidegger and
Augustine) at the same time. Death and eternity constitute the
polar limits of *narrated* time rather than phenomenological time.
Ricoeur links Heidegger and Augustine on the level of a philosophical anthropology built around the notion of narrative under-

standing. Human narrative self-identity is achieved, ultimately, by thinking about one's possible death in light of a perspective (eternity) that can imaginatively transcend death. Narrated time is the mediating third time between phenomenological time and cosmic time. The narrative world is constituted by the "interweaving references" of fiction and history. Fiction and history form a collaborative bridge between the phenomenological and the cosmological. In this interweaving collaboration, fiction stands closer to phenomenology, history stands closer to cosmology. Eternity and death, taken together, represent the limiting horizon for the collaboration of fiction and history in the constitution of narrative time. Eternity is the imaginative, fictive extension of phenomenology toward cosmology and the infinity that is implied in cosmic time. *Sein zum Tode,* on the other hand, represents a historical appropriation of the fact that cosmic time envelops us and determines our being. This historical appropriation is an extension of cosmology toward phenomenology. By focusing exclusively on a fact pertaining to our historicity (death), Heidegger overlooks the contribution that fictional imagination can make to self-understanding. Ricoeur extends Kant's understanding of the imagination and schematism beyond the confines of epistemology (Kant) *and* the confines of existential anthropology (Heidegger). Augustine and Heidegger are brought together on the basis of their different contributions to an anthropology grounded in a narrative understanding of human being in the world.

An imaginative category is never constitutive of hard, scientific knowledge. So Ricoeur's treatment of eternity is not based on any transcendental illusions. He is not claiming that eternity is some real, transcendent entity, only that it is a necessary aspect of our thinking. Eternity is one of the limits for our reflections on time and for our self-understanding. But what happens to the Eternal as a manifestation of God's own being when it is confronted with an argument such as Ricoeur's that takes its bearings from Kant's criticism of metaphysics? We who live after Kant cannot claim to "know" eternity as readily as Augustine could. Thus Ricoeur preserves the limits of reason as Kant conceived them. Like the ideas of reason, eternity can serve a regulative function—it can be a limit-concept—in our drive to understand ourselves and the world. But can it be more than that? Ricoeur's imaginative sense of eternity, almost as much as Heidegger's sense of death, seems to "occlude" the possibility of a specifically religious meaning. Religious meaning refers to the possibility of being confronted by a reality that is other than human consciousness and transcends it; that is, it seems to occlude the possibility that the divine, as Eternal *Verbum,* can point us toward the path of redemption from beyond the constraints of human temporality and finitude. I shall begin to formulate a solution to this problem as I now turn to the issue of Heidegger's influence on theologians.

Heidegger and the Theologians

In "Martin Heidegger and Marburg Theology," Hans-Georg Gadamer claims that the earliest form of *Being and Time* was an address Heidegger gave before the theological community of Marburg in 1924.[11] At that time, Heidegger said, " 'It is the true task of theology, which must be discovered once again, to seek the word that is able to call one to faith and preserve one in faith,' " (quoted in Gadamer, 198). Yet Heidegger's talk, according to Gadamer's recollection, "reflected a despair at the possibility of theology itself" (198). Even if Heidegger despaired of the possibilities for theology, some of the theologians in his audience did not.[12] One of Heidegger's colleagues at Marburg was Rudolph Bultmann. Perhaps more than any other theologian, Bultmann attempted to rethink the Christian message within the anthropological parameters (or existential "questions") that Heidegger established in *Being and Time.*

Bultmann's kerygmatic response to Heidegger's understanding of death is best depicted in "The Historicity of Man and Faith." This essay is comprised of two fairly independent parts, both of which fit nicely into Bultmann's overarching "demythologizing" project: to uncover, through interpretation, the understanding of existence inherent to the Christian message as we have received it in its mythological format. (In other words, Bultmann is asking about what sort of questions the Christian kerygma is capable of answering. He claims that it answers existential questions of the sort Heidegger raises.)

In the first part of "The Historicity of Man and Faith," Bultmann responds to Gerhard Kuhlmann's published misgivings about Bultmann's dependence upon Heidegger's existential analysis of human nature. At issue is the relationship between philosophy and theology. Kuhlmann is hesitant to allow that theological inquiry into the life of faith inevitably depends "upon a pretheological understanding of man that, as a rule, is determined by some philosophical tradition."[13] He accuses Bultmann of turning theology into "pseudophilosophy," or of profaning Christian revelation. Kuhlmann's concern is that if theology is always already dependent upon a philosophical preunderstanding of the full range of human possibilities, then it seems deprived of its own rightful ability to speak (in faith) of a revelatory Word of God. How can there be revelation if we already understand what can and cannot be possible for us as human beings? What can revelation add that is "new" or that is not already given in philosophy's privileged preunderstanding?

The response to Kuhlmann is built around Bultmann's use of Heidegger's distinction between the ontical and the ontological (or the existentiell and the existential). Philosophy is an ontological science that depicts on a formal level the conditions of possibility for concrete human existence. Or, to put it in Tillichian terms, philosophy depicts the range of possible questions that human beings can ask about their own existence. Theology, on the other hand, is an ontical or positive science that depicts on a concrete level how a particular human possibility—already formally understood by philosophy—can be actualized. Theology "depends" upon philosophy, or makes use of philosophy, by learning what it can about the phenomenon of human being in the world that philosophy discloses. Only when theology gets the questions right (via philosophy) can it proffer the Christian message as an answer for particular persons of faith.

The realm of ontological possibilities (philosophy) is a realm of questions; the realm of ontical actualities (theology) is a realm of answers. Both disciplines are concerned with the human as phenomenon: one formally, the other concretely; they need not be in competition. Thus Bultmann is able to contest Kuhlmann's objections:

> Theologically expressed, faith is not a new quality that inheres in the believer, but rather a possibility of man that must constantly be laid hold of anew because man only

> exists by constantly laying hold of his possibilities. The man
> of faith does not become an angel, but is *simul peccator,*
> *simul justus.* Therefore, all of the basic Christian concepts
> have a content that can be determined ontologically prior to
> faith and in a purely rational way. All theological concepts
> contain the understanding of being that belongs to man as
> such and by himself insofar as he exists at all. Thus theology
> should indeed learn from philosophy—precisely from that
> philosophy "which confesses as its deepest determination
> 'to serve the work of Dilthey,' which is to say, man's
> understanding of himself *qua* man." ("The Historicity of
> Man and Faith," 96)

The only thing more or new that the person of faith knows through
revelation is that he or she has actually been encountered by a
revelatory event that extends forgiveness, grace, and authenticity
to life *now*. "Every man, because he [already] knows about death,"
says Bultmann, "can also know about revelation and life, grace
and forgiveness" (100). Without a preunderstanding of our own
precarious situation, without an *interest* (or question) provoking
us to search for relief from death, how could we even care to hear
the Christian gospel?

In the second part of "The Historicity of Man and Faith,"
Bultmann reflects upon the concrete problem of our historicity
through a comparison of Gogarten and Heidegger. According to
Heidegger, what constitutes historicity—the sphere of possibilities
or choices that pertain to our immediate situation in life—is the
resolution that sees death as our most proper possibility. We ought
to choose or resolve upon possibilities for ourselves, or engage
ourselves factically in history, from the standpoint of our *Sein zum
Tode.* According to Gogarten, on the other hand, what constitutes
historicity is encounter with the neighbor as "thou." "For Heidegger
man is limited by death; for Gogarten he is limited by the thou"
(103). Bultmann's task, then, is to find a way to reconcile these two
apparently different points of view of the same phenomenon
(historicity). Can Heidegger's "resolution" be interpreted as Gogarten's "love"? (More specifically, can Heidegger's "others"—the
persons other than Dasein with whom Dasein is always already
involved and who become authentically visible because of Dasein's
resolve to be itself—be interpreted as Gogarten's "neighbors"?)

Bultmann's answer is yes, and he resorts to the relation between the ontological and the ontical to facilitate the reconciliation. But he conceives the ontical sense of love in a careful manner. Love is not simply an ontical *expression* of resolve, indicating a particular "what" upon which we might resolve. We do not resolve to love, but instead resolution as our most authentic possibility is itself *realized* in love. Love is the ontic determination of resolve, not an ontic expression of it. "Therefore," claims Bultmann, "if there is to be a unity between Heidegger and Gogarten, we must be permitted to say that only where the I decides for the thou, acknowledging his claim in love, is resolution actually realized" (105). Questions concerning the *realization* of possibilities are ontical and not ontological. They concern factual experience (history).

Since Heidegger speaks as an ontologist, he has no right to speak of (ontical) love, according to Bultmann. Gogarten's concern is theological (ontical), hence he wants to demonstrate that actual history only occurs where the I hears the claim of the thou. But in order to understand what is at stake in Gogarten's theological work, we must be clear on the formal/ontological level about what the basis for human historicity is. Heidegger's ontology teaches us that our historicity pertains to the realization of possibilities; in fact, Dasein's own being must be defined as a possibility of being. Among other things, that means the potential to be either inauthentic or authentic is a constitutive possibility for Dasein. As a possibility of being, Dasein can be genuine or not, for one of Dasein's "historical" options is not to pursue its ownmost possibility. Only in light of this ontological discussion of authenticity can the significance of Gogarten's theology be understood: We are only genuinely (authentically) historical in love; (ontical) love is the only possible determination of authentic (ontological) resolve. Theology claims that only as a Christian is the human genuinely historical. Such a claim can be understood only in light of ontological clarification of the possibilities of fallenness and authenticity. "Love is not *caritas infusia*," says Bultmann, "but rather is from the outset an ontological possibility of human existence of which man dimly knows" (108).

Bultmann's final point is that, for the person of faith who stands in love, *Sein zum Tode* is no longer the limit constituting human historicity.

> For him who knows himself loved, however, it becomes
> clear that the actual limitation of the I is given by the thou,
> and death forthwith loses its character as the limit. The
> question concerning death becomes superfluous for him
> who knows (existentiell) that he is there to serve the
> neighbor. (108)

Bultmann is suggesting that the thou can displace death as
limit-idea in the concrete situation of genuine history. In the
circumstance of faith, "love is an absolute surrender of the I and
. . . as such 'overcomes' death" (110).

Despite working within the constraints of Heideggerian philosophy, Bultmann does not sacrifice religious meaning to ontological death. Rather, he finds a way to articulate the Christian
message of love in terms of Heideggerian insight. The message of
love "answers" the question of authentic resolve in the face of
death as its realization in a concrete act of faith. But Bultmann
never questions the "questions" that Heidegger's philosophy
thrusts upon him because he is solely concerned with reinterpreta-
tion of the kerygma. Theology does not raise questions about
human existence; that is philosophy's task. Instead, it interprets
the Christian message in light of the preunderstanding that
philosophy gives. Only the answering side of Tillich's method of
correlation is operative in Bultmann's construal of theology based
on Heidegger's *Being and Time.*

Another of Heidegger's theological colleagues at Marburg was
Paul Tillich. Unlike Bultmann, however, Tillich was influenced
by Heidegger in nonspecified ways. He never offers a sustained
interpretation of Heidegger's work (though he does insist that
Being and Time represents a "theonomous philosophy," which I
address later), but Tillich's discussion of ontological finitude (and
existential estrangement) is informed by *Being and Time.* Also,
Tillich's method of correlation both expresses Heideggerian
insights and enables him to interpret Heidegger's philosophy in its
"theonomous" significance. The notion of *Sein zum Tode* in no
way precludes Tillich's discussion of the "essentialization process" through which human beings are incorporated into the
eternal life of God.

In his *Systematic Theology,*[14] Tillich claims that theology has
always used a method of correlation that answers existential

questions by referring to the Christian message. He explains the basic insight that gives rise to the correlation accordingly:

> Whenever man has looked at his world, he has found himself in it as a part of it. But he also has realized that he is a stranger in the world of objects, unable to penetrate it beyond a certain level of scientific analysis. And then he has become aware of the fact that he himself is the door to the deeper levels of reality, that in his own existence he has the only possible approach to existence itself . . . Whoever has penetrated into the nature of his own finitude can find the traces of finitude in everything that exists. And he can ask the question implied in his finitude as the question implied in finitude universally. (I, 62–63)

Like Heidegger (and Augustine), Tillich is suggesting that ultimate reality manifests itself in the life of human beings. Dasein is the "there" of being itself; the soul participates in the being of God. Tillich's correlation is based upon the insight that we already have a relationship with, and an understanding of, the ultimate. Theology, like Heidegger's philosophy, is a hermeneutic of human being in its fragility and its lack of authenticity. It teaches us to follow the questions that arise from our sense of shock and anxiety at just being alive to their conclusion (which is also their beginning) in a religious quest for the transcendent. Then the Christian symbols can disclose the reality of God as both the ground of being and the ultimate concern of human life. This "answer" is already implied or partially given in our existential questions themselves since we could never ask about what is totally unfamiliar. Tillich claims that the religious quest is fundamental to human life and that it is provoked by an experience of the holy.

Tillich's perspective depicts what he calls the Augustinian or "ontological type" of philosophy of religion, which he contrasts with a "cosmological type."[15] The ontological perspective assumes that a preunderstanding of God, based upon primordial experience, is what gives rise to the theological question of God. In other words, we would never search for what we did not already know in some sense. When Tillich claims that *Being and Time* must be read as a "theonomous philosophy," he is suggesting that Heidegger is essentially an Augustinian thinker.[16] Dasein already understands being itself in a primitive way, for that is what provokes Dasein to ask the question of being. By pursuing that fundamental question, Dasein can come to grips with its own finite and fallen nature. Thus Tillich's claim that Heidegger's style of thinking parallels theological thinking in its Augustinian form. Hence also Tillich's willingness to be influenced and informed by the existential philosophy that he was first introduced to when listening to Heidegger at Marburg.[17]

By focusing on the method of correlation as such (which includes questions *and* answers), Tillich is able both to appropriate Heideggerian insights and to place Heidegger's thought itself within a theological context. His interaction with Heidegger represents an advance beyond Bultmann because theological correlation has become the issue. Only this way can the theological significance of Heidegger's thought be truly assessed. And Tillich's claim is that Heidegger's thought is a theonomous philosophy.

Tillich's *Systematic Theology* is divided into five parts. In the individual parts, which represent the structural spheres of existence (rational, ontological, existential, living unity of ontological and existential, and historical), the correlation of philosophical question and religious/symbolic answer is employed: Revelation answers the ambiguities of reason; God answers the questions implied in finitude as such; Jesus as the Christ answers the questions that grow out of existential estrangement; Spirit answers the questions implied in the ambiguities of life; and Kingdom of God answers the questions that arise from our experience of history. Tillich's discussion of time, death, and eternity appears in the fifth part of the system, entitled "History and the Kingdom of God."

This historical part of Tillich's system grows out of the fourth part, which deals with life as the ambiguous unity of the essential and existential (or fallen) aspects of existence. History constitutes the temporal dimensions of life, depicting its whence and whither. Eschatological questions about the ultimate fulfillment or end of history are answered by the symbol of "Eternal Life." At issue is the relation of time and eternity. Whereas the symbol of creation depicts a past transition from eternity to time, the eschatological symbol depicts a future transition from time back to eternity. "God *has* created the world," says Tillich, "and he *will* bring the world to its end" (III, 395). This is reminiscent of Augustine's implied claim that there are two senses of time (creative and redemptive) that derive from two possible perspectives toward the eternal (as source or fulfillment of time). Our orientation toward the end of history is determined by a present experience of the eternal as the "now" in which past and present (creation and redemption) come together. But what does Tillich mean by suggesting that the temporal is transposed into the eternal? How does this transition occur?

Tillich claims that in the transition from the temporal to the eternal "the negative is negated." Being is positive, nonbeing is negative. Life is an ambiguous mixture of the two. The fulfillment of life—the attainment of eternal life—is a consequence of God's judgment, which liberates being from nonbeing. God's "eternal memory" only acknowledges the positive, never the negative. Tillich calls this process of fulfillment "essentialization." It entails the incorporation of two "positives" into the eternal life of God. These two positives are a person's (or a thing's) essential being (given in the original transition from eternity to time) and its new being, which stems from within existence. The ambiguities of life, which threaten to disrupt the self-world structure of existence in its various polar constituents, are completely conquered in the process of essentialization; the functions of spirit (morality, culture, religion) are also completely fulfilled. Essentialization, or the elevation of the positive into eternal life, is the actualization of God as love.

Tillich speaks of an "eternal death" that bears some resemblance to what I have called "ontological death." Eternal death refers to a breakdown in the essentialization process. The temporal cannot reach the eternal, it cannot transcend its own finitude. I have suggested that such a loss of transcendence is implied in Heidegger's *Sein zum Tode*. The eternal is permanently at bay. Yet Tillich's discussion concerns the theological problem of eternal damnation versus universal salvation. Essentializaton seems to imply that everyone will be saved as a matter of course, which might compromise the seriousness of life and its ethical choices. Tillich gets around this problem by claiming that eternity is both like time (it is not timeless identity) and different from time (it is not permanent change). That is, the life process is continued in eternity—eternal life is still life—so that what is essential and positive in an individual can continue to be developed. The person who has lived a bad life will not be able to participate as fully in eternal life since that person will have fewer positive qualities to essentialize. There are degrees or levels of eternal life, according to Tillich; hence, the seriousness of life can be preserved despite the fact that all beings are taken up into the divine life.

Tillich's understanding of time and eternity is Augustinian, though he differs from Augustine on one point: He is more concerned with the "becoming" that is implied in speaking of eternal or divine life than was Augustine. For Tillich, eternity is not altogether discontinuous from the temporal process. He claims that Augustine conceived time in terms of a straight line (he does not say where Augustine argues this) rather than a circle, as Greek philosophers did. Time runs from its beginning in creation to its end in redemption.

> However, the diagram of the straight line does not indicate the character of time as coming from and going to the eternal. And its failure to do so made it possible for modern progressivism, naturalistic or idealistic, to prolong the temporal line indefinitely in both directions, denying a beginning and an end, thus radically cutting off the temporal process from eternity. This drives us to the question as to whether we can imagine a diagram which in some way unites the qualities of "coming from," "going ahead," and "rising to." I would suggest a curve which comes from above, moves down as well as ahead, reaches the deepest point which is the *nunc existentiale,* the "existential now," and returns in an analogous way to that from which it came, going ahead as well as going up. This curve can be drawn in every moment of experienced time, and it can also be seen as the diagram for temporality as a whole. (III, 420)

Tillich's "curved" understanding of time and eternity modifies and clarifies Augustine's understanding. So, despite Heidegger's philosophy of finitude, and the jeopardy to religious meaning implied in *Sein zum Tode* as terminus of critical thought, Tillich (like Bultmann and other theologians) still speaks of the Christian symbol of Eternal Life. *Being and Time,* from a Tillichian perspective, enables us to understand our ontological shock in greater depth, which calls all the more for a religious answer. According to Tillich, the rise of existential thought is Christianity's good fortune.

Ricoeur's argument in *Time and Narrative* stands in line with Tillich's theology in that it employs the method of correlation. The philosophical questions raised about time can be answered only in correlation with narrative constructions that interweave the isolated poles of our thinking. Ricoeur's confrontation between Heidegger and Augustine is also conceived in terms of correlation. Heidegger on death is correlated with Augustine on eternity: Death is conceived as the historical limit of human temporality, eternity as the corresponding poetic limit. Like Bultmann and Tillich before him (both classic theologians influenced by Heidegger's *Being and Time*), Ricoeur attempts to limit the ultimacy of *Sein zum Tode* for human temporality and self-understanding by confronting (or "answering") it with some sense of eternity (whether it is Christian love, eternal life, or imaginative extension of thought).

I wish to make two points in placing Ricoeur within the company of Bultmann and Tillich. The first is that Ricoeur extends the hermeneutical horizon for employment of the method of correlation well beyond both Bultmann and Tillich. Bultmann's "answer" to Heidegger's *Sein zum Tode* is given in terms of the preached kerygma (the Christian message of love) that enables the listener to become resolute and authentic. Tillich's "answer" is given in terms of the Christian symbol of Eternal Life that, in pointing to the reality of God's life, already participates in it (thus initiating an essentialization process). Notice that both kerygma and symbol, the forms of the answers given by Bultmann and Tillich, respectively, refer to linguistic experiences. Language is the medium of disclosure for human being in the world. The problem with kerygma and symbol is that they only partially express the full power of language. As we shall see in the next chapter, Ricoeur claims that the full disclosive powers of language are expressed in texts. So the method of correlation is extended beyond Bultmann and Tillich by Ricoeur's textual theory of hermeneutics. Ricoeur "answers" Heidegger in the context of a textual correlation: *Being and Time* is confronted with the *Confessions* of Augustine. A promise of better understanding is enkindled by the advance from kerygma and symbol to the level of the text. Bultmann's demythologized kerygma is artificially isolated from its textual meaning, and Tillich's religious symbols are confined to a level of discourse (word or sentence) below that of texts. By following Ricoeur's textual correlation beyond Bultmann and Tillich, we can (potentially, at least) advance our theological understanding of Heidegger and Augustine.

However, there is a glitch in this scenario for the advancement of understanding. It concerns the second point I wish to make about placing Ricoeur in the company of Bultmann and Tillich. The point is that Ricoeur's textual correlation fails to advance theological understanding beyond Bultmann and Tillich because Ricoeur neglects the religious sense of Augustine's eternity. He fails to take into account the textual significance of the Eternal *Verbum*—the redemptive voice of God—in the *Confessions*. Instead, he interprets eternity as an imaginative idea that only extends our thinking beyond existential limits. Ricoeur's work holds much promise for theology, but he does not quite fulfill it.

"The Being of God When God Is Not Being God"

As I demonstrated earlier, Ricoeur's treatment of Heidegger and Augustine can be placed within a theological context. The issue, theologically speaking, is this: Can we still understand and appreciate eternity as an aspect of God's own being that both transcends time and fulfills it in light of the increased ontological significance of death that is best depicted in Heidegger's notion of *Sein zum Tode?* Bultmann and Tillich both answered this question in the affirmative. Ricoeur's textual hermeneutics advances the potential for understanding the correlation of death and eternity (or philosophy and religion). Following Scharlemann, I fulfill that potential.

In "The Being of God When God Is Not Being God," Scharlemann speaks to the issue of Heidegger and the theologians. He indicates how theologians, following Heidegger's example, can still work toward accomplishing a destruction (or deconstruction) of the metaphysically driven "theistic picture" that continues to dominate Western thought. He claims that the kind of destructive rereading of the philosophical tradition that Heidegger accomplished in *Being and Time* has yet to be duplicated for the theological tradition. After giving a brief review of Heidegger's destruction of the history of ontology, Scharlemann turns to the theological tradition. He argues that the otherness of God has remained unthought and conceptually forgotten in theology, just as, according to Heidegger, the question of the meaning of being has remained unthought and conceptually forgotten in philosophy.

To read the history of thought deconstructively, as Heidegger did, means to read it backward until the experience that spawned traditional thinking can be recalled. Deconstruction "takes the tradition back to earlier concepts . . . until it reaches the point where the concepts are first formed as concepts."[18] Heidegger turned to this genealogical style of interpretation in order to combat a particular problem of forgetfulness inherent to the tradition of thinking in the West. According to Heidegger, we have forgotten the significance of the question of being—the originating question in the history of thought—because the concepts and tradition engendered by that question have displaced its personal, experiential basis. Hence, we must read through traditional concepts and debates in order to recognize again the role that being has played (and continues to play) in our lives and thinking.

Specifically, Heidegger attempted to rethink the question of the meaning of being in its originary significance by destructing or following back the history of ontology. Ontology pertains to the issue of how time and being are linked within the existent being of Dasein. He traced the issue of ontology back to the Greeks, who did understand being on the basis of time originally: The meaning of being was understood in terms of the present. But this original understanding of time in its co-implication with being—the being of entities in the world is brought into view or made present through time—was contradicted when the Greeks conceived both being and time as entities among other entities. So the Greeks distorted an original understanding when they conceptualized it. "In these concepts," states Scharlemann, "being and time appear as entities that, nonetheless, are not entities; they involve a contradiction in the concepts defining them because the difference between time and other entities, like the difference between being as such and entities, remained unthought" (86). Deconstruction aims to think back to the point in the history of thought where a self-contradiction was manifested, and then to overcome the contradiction by thinking what was left unthought. The assumption is that if we are able to think what was left unthought, then we will have recovered the original understanding of experience that was both captured and contradicted in the initial conception.[19]

Corresponding to the oblivion of the meaning of being, claims Scharlemann, there is a forgetting of the otherness of God. "The symbol of the otherness of God (incarnate deity, or existent deity) is subject to the same oblivion in the history of theology as the question of the meaning of being is in the metaphysical tradition" (88). Just as Heidegger wanted to rethink how being is present in time and in the world, Scharlemann wants to rethink how the Creator is present in creation. He wonders how we can think of God "existing not as a transtemporal or metaphysical entity but as an actuality in life and history" (107). So the issue of being and time in philosophy correlates with the issue of Creator and creation in theology. But the doctrine of creation has engendered a "theistic picture" where creature and Creator are thought of as separate entities that come together (barely) only in a logical conception of the whole as being. However, what cannot be thought in the tradition of this picture, claims Scharlemann,

> is that the world is itself a moment in the being of God; what cannot be thought is that the world is the being of God when God is not being deity, or the being of God in the time of not being. To reconstruct the picture of the relation between God and the world, after this destruction of the picture, requires rethinking the division between the uncreated and the created according to the idea of God's being God as God and God's being God as other than God. (90)

Scharlemann illustrates the self-contradictory nature of the theistic picture—namely, its tendency to cover up and leave unthought a more basic understanding of God's being in the world—by analyzing Anselm's definition of God.

Augustine's discussion of time and eternity, as principal figures for creature and Creator, is open to Scharlemann's kind of deconstructive reading. The self-contradiction inherent to Augustine's analysis, which indicates its primordiality or proximity to religious experience, is entailed in his attempt to isolate a sustained reflection on time from a meditation on eternity. That is,

Augustine gives in to the tendency within the traditional theistic picture to separate creature and Creator. But this separation is self-contradictory in Augustine's case because of what he says and implies about the relation between time and eternity. Augustine says that both past and future have their beginning and their end in eternity (*Confessions,* ch. 11, 262). This implies that God is more than just a maker of time, an uncreated being separated from the created world. It implies, in fact, that God is present in time as both its source and its "coordinator." What remains unthought in Augustine's analysis is the case where time is, in some sense, the being of God when God is not being God. To be more precise, Augustine leaves unthought how the creative Word becomes incarnate Word within time itself. For Augustine, the Eternal Word is both Creator and Redeemer. But in this "equivocation," he does not explore how one becomes the other, how the time of creation becomes redemptive when the Redeemer enters into the torn and distended temporal soul as a guiding light and healing force. Redemptive time is the being of God (as Eternal Word) when God is not being God. In Augustine's text, however, the symbol of the otherness of God, of God's worldly presence, is overlooked.

Notes

1. Pöggeler, *Martin Heidegger's Path of Thinking,* 47.
2. "Basically, all ontology, no matter how rich and firmly compacted a system of categories it has at its disposal, remains blind and perverted from its ownmost aim, if it has not first adequately clarified the meaning of Being, and conceived this clarification as its fundamental task" (*Being and Time,* 31 [11]). Note that I give page references to the German edition of this text, *Sein und Zeit,* in parentheses (or brackets) following references to the English edition.
3. Perhaps it would be helpful to explain why Heidegger bothers to make this somewhat puzzling distinction between the ontical and the ontological (or the existentiell and the existential). Several times he makes the point that "the entity which in every case we ourselves are, is ontologically that which is farthest" (359 [311]). Ordinarily, ontically, we are lost in self-deception. Hence, any "ontology" that begins its inquiry with Dasein as it is ordinarily given or understood (as all ontologies prior to Heidegger's fundamental ontology did) will fall into the trap of furthering Dasein's self-deception. That is, most ontologies have moved from the ontical to the ontological, thus overlooking (and perpetuating) Dasein's ontical fallenness. Heidegger reverses the process. He first uncovers Dasein's ontical fallenness through ontological analysis, then presents an authentic vision of factical Dasein. "The laying-bare of Dasein's primordial Being must rather be wrested from Dasein by following the opposite course from that taken by the falling ontico-ontological tendency of interpretation" (359 [311]). The distinction between the ontical and the ontological is necessary because of Dasein's fallen nature, which manifests itself in misinterpretations.
4. Heidegger says: "Because discourse is constitutive for the Being of the 'there' (that is, for states-of-mind and understanding), while 'Dasein' means Being-in-the-world, Dasein as discursive Being-in, has already expressed itself. Dasein has language" (208 [165]).
5. Following Gerald Bruns's interpretation of section 32 of *Being and Time* ("On the Weakness of Language in the Human Sciences," in

The Rhetoric of the Human Sciences [Madison: University of
Wisconsin Press, 1987], 239–62), where Heidegger discusses the
nature of understanding, it is possible to see how much Ricoeur's
argument in *Time and Narrative* depends upon Heidegger. Bruns
emphasizes the imaginative aspects of Heidegger's interpretation of
understanding. In that sense, Heidegger's "hermeneutical 'as' "
anticipates Ricoeur's imaginatively variable mimetic constructs.
Both Heidegger and Ricoeur rely upon and extend Kant's understanding of the productive imagination. See "Beginning a Theological Extension" in my next chapter for further discussion of this
point.
6. "Idle talk, curiosity and ambiguity characterize the way in which, in
an everyday manner, Dasein is its 'there'—the disclosedness of
Being-in-the-world. As definite existential characteristics, these are
not present-at-hand in Dasein, but help to make up its Being. In
these, and in the way they are interconnected in their Being, there is
revealed a basic kind of Being which belongs to everydayness; we
call this the 'falling' of Dasein. . . . In falling, Dasein *itself* as factical
Being-in-the-world, is something *from* which it has already fallen
away" (220, 221 [176]).
7. "If the Interpretation of Dasein's Being is to become primordial, as
a foundation for working out the basic question of ontology, then it
must first have brought to light existentially the Being of Dasein in
its possibilities of *authenticity* and *totality*" (276 [233]).
8. See section 32 of *Being and Time,* "Understanding and Interpretation," for Heidegger's discussion of meaning as the "upon which"
for any projection of Dasein's possibilities.
9. See, for example, Cassirer's "Kant and the Problem of Metaphysics:
Remarks on Martin Heidegger's Interpretation of Kant." Cassirer
praises Heidegger for fighting the neo-Kantian trend to assume that
Kant's essential goal was to ground metaphysics in a scientific
epistemology. The question that motivated Kant, according to
Heidegger, is none other than the question of fundamental ontology:
What is the human as an opening for and to being? Yet, in the end,
Cassirer accuses Heidegger of reading his own philosophy into
Kant's text.
10. *Kant and the Problem of Metaphysics,* 3.
11. Gadamer, "Martin Heidegger and Marburg Theology," 199.
12. I am not attempting to review the total impact of Heidegger's
thought on theology. There can be no doubt that it is a considerable
influence, but it is beyond the scope of my interests to investigate
that question. I focus on specific theologians who have been
influenced by Heidegger in order to make one point: From the

beginning, theologians have responded to the threat to religious meaning entailed in Heidegger's notion of *Sein zum Tode*. So my attempt to respond to Heidegger with a reading of the *Confessions* is not unusual. It does fit within a theological tradition. Since my response to Heidegger is text specific—confronting *Being and Time* with the *Confessions*—it is also beyond the scope of my interests to investigate the *Kehre* (or turn) in Heidegger's thought and the different possibilities for theology opened by the later works of Heidegger. Nor do I comment on the influence that theologians such as Schleiermacher and Kierkegaard had on Heidegger's thought. Again, my task is to offer a conjoint reading of two texts, not to trace the mutual influences of theologians upon philosophers, or vice versa.

13. Bultmann, "The Historicity of Man and Faith," 98.
14. Tillich's *Systematic Theology* is a three-volume work. The roman numerals preceding the page numbers of quotations from this work indicate the volume from which I am citing.
15. See Tillich, "The Two Types of Philosophy of Religion," in *Theology of Culture,* 10–29.
16. See Tillich, *The Interpretation of History,* 39–40. Note that I have relied upon Thomas O'Meara ("Tillich and Heidegger: A Structural Relationship") as a guide to Tillich's citations of *Being and Time.*
17. See Tillich's "Autobiographical Reflections," in *The Theology of Paul Tillich,* 14.
18. Scharlemann, "The Being of God When God Is Not Being God," 85.
19. "Destruction analyzes the history of thought in order to discover what the thought is about by getting back to its origins, which appear at the point of self-contradiction in the thought" (ibid, 99).

Chapter 4
Time and Narrative:
A Theological Extension

Eternal Truth, true Love, beloved Eternity—all this, my God,
you are, and it is to you that I sigh by night and day.
                    Confessions, Book VII, ch. 10, p. 147

Ricoeur said that the most serious question he asks in *Time and Narrative* is how it is possible to think about death and eternity at the same time. His answer is determined by a rather nonexplicit confrontation between two texts: Augustine's *Confessions* and Heidegger's *Being and Time.* In his criticisms of Heidegger's sense of *Sein zum Tode,* Ricoeur creates a place in his theory of time for an interpretation of Augustine's sense of eternity as the other of time. Ricoeur argues that Heidegger's notion of *Sein zum Tode* is derived from a false sense of unity between historical fact (since we are all subject to cosmic time, we all must die) and imaginative projection (since we all create a sense of temporal unity that includes references to future and past, eternity is no less inconceivable than is death). Ricoeur's analysis separates the cosmological and phenomenological strands of thought that Heidegger collapsed. That opens the door for retrieval of a sense of eternity in the context of narrated time, whereby eternity is linguistically construed as a limiting horizon for the imaginative projections that can extend phenomenological thinking. Death and eternity are brought together as the twin limiting horizons of a new, third sense of time—narrated time—that poetically interweaves phenomenological time with cosmic time. Augustine's eternity is the pole within narrated time that stands closest to phenomenological time; Heidegger's ontological death is the pole within narrated time that stands closest to cosmic time. But what

of the religious sense of eternity in Augustine's text? Ricoeur does
not adequately deal with the religious and theological issues
implied in his confrontation of Heidegger and Augustine. He
reduces Augustine's sense of the Eternal as a manifestation of the
living God to little more than a Kantial limit-idea. Such curtail-
ment of the religious is more than critical thought requires.

I have yet to speak of Ricoeur's "Conclusion" to *Time and
Narrative,* but I do so in the first section of this chapter. The
deconstructive interpretation of the *Confessions* that Ricoeur
initiates in the "Conclusion" opens the door for the second step of
my strategy. I take that second step by turning to Scharlemann's
"The Textuality of Texts." I follow his suggestions for religious
and theological extension of Ricoeur's hermeneutics and, specifically, his *Time and Narrative.* His claim is that Ricoeur's theory
of text lacks a notion of "textuality" that would enable him to
understand the full meaning of religious texts. In the third section,
I once again take up the matter of Ricoeur's interpretation of the
Confessions. By extending Ricoeur's reading theologically—I
argue that Augustine's theory of time indicates that he thinks the
narrated time of redemption is a religious symbol, the being of
God when God is not being God—I uncover evidence of Scharlemann's concept of "textuality" in the *Confessions.* I also argue
that to the extent we can still hear the voice of textuality in
Augustine's text, and understand what it means symbolically, we
need not give up the full religious sense of eternity as an aspect of
God's being. Ricoeur's mediation between death and eternity is
sorely lacking from a theological point of view. Turning to
Scharlemann for help in extending Ricoeur's theory of text and his
reading of the *Confessions* remedies that situation.

"Conclusion" to *Time and Narrative*

The "Conclusion" is no simple summation of prior arguments.
Rather, it is a reflection on the limits of Ricoeur's principal
hypothesis that the poetics of narrative can offer a productive
response to the aporias that arise whenever we reflect upon the
nature of time. As Ricoeur rereads his own discussion of the
aporetics of temporality, he discerns three different problems or
levels of aporia that indicate the extent to which narrative can

provide an adequate response. Ricoeur thinks that the turn to narrative constitutes an adequate response only on the first level, where the aporia involves a tension between the phenomenological and cosmological perspectives on time. But a second level of aporia (he calls it "the problem of the totality of time") is concealed in and surpasses the split between the phenomenological and cosmological perspectives on time. Ricoeur thinks that the poetics of narrative is less able to reply to the aporetics of time on this level. Hence, a premonition of the limits of the working hypothesis (to confront the aporetics of time with the poetics of narrative) begins to appear. But there is a further, even more intractable aporia concealed behind the first two. On this third level, the ultimate unrepresentability of time—its mysteriousness—becomes apparent. Here, the limitations of Ricoeur's working hypothesis become most apparent.

With the recognition that a sustained reflection on time (poetic or otherwise) ultimately must give way to the mystery that time is, *Time and Narrative* opens itself to theological extension. This is indicated by the fact that Ricoeur actually returns to his starting point: He began with an interpretation of Augustine's *Confessions,* noting then the role that the mystery of eternity played in his reflections on time, and he ends with a "confession" of his own that imitates Augustine's prayers to the Eternal for help in understanding.

In the last section of the "Conclusion" to *Time and Narrative* ("The Aporia of the Inscrutability of Time and the Limits of Narrative"), Ricoeur says,

> My rereading reaches the point where our meditation on time not only suffers from its inability to go beyond the bifurcation into phenomenology and cosmology, or even its difficulty in giving a meaning to the totality that is made and unmade across the exchanges between coming-towards, having-been, and being-present—but suffers, quite simply, from not really being able to think time. (III, 261)

Ricoeur finds that time is finally "inscrutable," just as he found (following Kant) in previous work that the origin of evil is inscrutable.[1] Ricoeur also finds that, just as thinking can never master or exhaust the meaning of what it encounters in the enigma of time, narrativity can never exhaust or master the corresponding encounter with its own limitations to refigure time. To the extent that time is inscrutable, the response of narrative to the aporias of time is limited.

Ricoeur returns briefly to Aristotle, Augustine, Kant, Husserl, and Heidegger in order to expose an inscrutability factor in each of their reflections on time. He claims that, for Aristotle and Augustine, the mysterious or inscrutable manifests itself in the mythic archaisms—Greek and biblical, respectively—that emerge from within their philosophical speculations. Thus, Ricoeur's final contrast between Aristotle and Augustine is theologically motivated. Aristotle's Greek religious heritage is different from Augustine's Hebrew heritage.

Ricoeur claims that the tinge of archaism in Aristotle's philosophy can be discerned in his interpretation of "being in time" (*Physics,* Book IV, ch. 12). In part, "the 'in' expresses the precedence of time as regards any thinking that wants to circumscribe its meaning, hence to envelop it" (*Time and Narrative,* III, 262). The only way to speak of this already given time that envelops us—where it comes from—is to resort to myths about the genesis of time. It is possible, argues Ricoeur, to trace Aristotle's remarks back to Plato's "philosophical story" about time (*Timaeus*) and even farther back. Even though philosophy distanced itself from its mythic origins, persistent figures of the inscrutability of time—all born from mythic theogonies—betray an underlying connection. Mythic archaisms persist in philosophical reflection as figures of inscrutability. These figures are bound to the representation of something beyond time (eternity, divinity) that can help to evaluate human time.

Ricoeur insists that we can hear the second, Hebrew archaism of Western thought in Augustine's phenomenology. What points to the archaic in Book XI of the *Confessions* is the contrast between time and eternity that envelops the examination of time. Recall that in Ricoeur's first reading of the *Confessions (Time and Narrative,* Part I), he isolated three levels at which eternity affects the speculation about time. Here, in the "Conclusion," he reinterprets the three (eternity as limit-idea, as intensifying the experience of distension, and as model for time to emulate) so that they fit into a Hebraic scheme of praise, lamentation, and hope.

> It is first in a spirit of praise that Augustine celebrates the eternity of the Word that remains when our words pass

away. So immutability plays the role of a limit-idea with regard to temporal experience marked by the sign of the transitory. Eternity is "always stable"; created things never are. To think of a present without a future or a past is, by way of contrast, to think of time itself as lacking something in relation to this plenitude; in short, as surrounded by nothingness. Next is is in the mode of lamentation, within the horizon of stable eternity, that the Augustinian soul finds itself exiled to the "region of dissimilarity." The moanings of the lacerated soul are indivisibly those of the creature as such and the sinner. In this way, Christian consciousness takes into account the great elegy that crosses cultural frontiers and sings in a minor key about the sorrow of the finite. And, finally, it is with a note of hope that the Augustinian soul traverses levels of temporalization that are always less "distended" and more "firmly held," bearing witness that eternity can affect the interior of temporal experience, hierarchizing it into levels, and thereby deepening it rather than abolishing it. (III, 264–65)

Despite the Platonic, speculative tendencies in Augustine's own thought, Ricoeur thinks that we can still hear a specifically Hebraic form of speaking behind the praise, lamentation, and hope that accompany Augustine's speculation on eternity and time. Here, Ricoeur makes explicit the deconstructive tendency in his second reading of Augustine that I noted earlier (*Time and Narrative,* Part IV).[2] The principal effect of this archaism upon Augustine's thought is to establish an inherent resistance to the Hellenization of every sense of eternity. Exegesis of this Hebraic form of speaking, says Ricoeur, "reveals a multiplicity of significations that prevent eternity from being reduced to the immutability of a stable present" (III, 265). The most precious Hebrew sense of eternity is given in the idea of God's fidelity to the people of the Covenant.[3] Ricoeur's text is rather cryptic at this point. He is interested only to suggest that as much is concealed in Augustine's notion of the eternal present as is revealed. Ricoeur never follows through with the deconstruction that he initiates. As a philosopher first, he only alludes to a theological level for his inquiry. It is our task to push Ricoeur further in the theological direction that he himself points out.

What, then, are the limits of narrativity relative to the inscrutability of time? How can narratives refigure a time that is unrepresentable because it has always already enveloped us in its own mystery whenever we attempt to construe it? Ricoeur considers two different kinds of limit in this circumstance, one internal to narration, the other external. For my purposes, only the limit internal to narration merits discussion. Within their own domain, narratives—especially fictional narratives—exhaust themselves in attempting to draw near the inscrutable. Through imaginative variations, the three fictional tales about time that Ricoeur investigated were able to present limit-experiences that pertained to eternity, or time's "other." Fiction is able to multiply our experiences of eternity, says Ricoeur,

> thereby bringing narrative in different ways to its own limits. This multiplication of limit-experiences should not surprise us, if we keep in mind the fact that each work of fiction unfolds its own world. In each instance, it is in a different possible world that time allows itself to be surpassed by eternity. This is how tales about time become tales about time and its other. (III, 271)

And in staking out the borderlines of eternity, "the limit-experiences depicted by fiction also explore another boundary, that of the borderline between story and myth" (III, 271). Here we see a basic pattern in Ricoeur's thought repeating itself: A reflection on time (like a reflection on evil) makes an inevitable return to its mythic/religious sources.

Beginning a Theological Extension

The first step in my strategy to offer a further theological response to ontological death is now complete. I have followed Ricoeur's argument in *Time and Narrative* from its initial philosophical juxtaposition of Augustine and Aristotle to its final theological juxtaposition of them. Now it is time to take the second step, which entails the development of Ricoeur's own thought even while moving beyond it. I begin this second, theological step by reviewing Scharlemann's "The Textuality of Texts." I turn to Scharlemann because he offers a reading of *Time and Narrative* that suggests theological extension. Scharlemann's reading of Ricoeur is critical, and yet the criticism attempts to develop a tendency already implied in Ricoeur's work from its earliest conceptions.

"The Textuality of Texts"

In "The Textuality of Texts," Scharlemann argues that Ricoeur's hermeneutics is insufficient in two respects. First, Ricoeur's hermeneutics does not develop an adequate definition of texts, one in which the symbolic phenomenon of textuality can come into view. "Textuality" refers to the unique mode of being that texts have. Having textuality means that texts are alive in some sense; they are embodied voices that can confront us with something as unique and different as any embodied person that we might meet. Scharlemann does not think Ricoeur's definition of text as "any discourse fixed in writing" allows for the appearance of textuality.[4]

The second insufficiency that Scharlemann notes in Ricoeur's hermeneutics concerns the kind of world that literary texts can open up for their readers. He thinks that Ricoeur's understanding of a text's world is determined solely by contemporary historical consciousness. Ricoeur "does not ask whether there is a consciousness, perhaps other than historical, for which the relation of future and past is different from what it is in existential and historical time" (14). According to Scharlemann, biblical resurrection narratives (as well as other texts that refer to the resurrection) are not understandable within the historicist perspective implied in Ricoeur's textual hermeneutics.[5] Resurrection narratives present a being beyond death that counterread the *Sein zum Tode* of our existential world. How can a biblical text present such an alternative world? That is Scharlemann's question. He concludes that there must be something more to texts than Ricoeur's hermeneutics seems to allow. Hence the notion of "textuality":

> A text's world is not just a redescription of the real or existential world [as Ricoeur sees it] but, we can say, a being there of another world (a "worlding" of the world, in Heideggerian language) and another voice just as each human person is a being-there of the self in the world of concern; and, if that is the case, then the textuality of a text is

an articulated body which localizes a being in the world just
as the human flesh and bones are an articulated body which
localizes, or gives a here-and-now to, a being in the world.
(15)

In the first part of "The Textuality of Texts," Scharlemann
indicates that *Time and Narrative* repeats a predictable pattern
that is twofold. This basic pattern, first encountered in Ricoeur's
reflections on evil in *Philosophy of the Will,* is to move from
theoretical reflection on some problem (e.g., the possibility of
evil) to hermeneutical interpretation of the myths and symbols
that gave rise to the theoretical problem in the first place (the fact
of evil).[6] The second aspect of the pattern in Ricoeur's thought has
to do with his hermeneutical theory, which is based upon his
understanding of the metaphoric process. Scharlemann attends to
the second aspect of Ricoeur's pattern of thought first:

> There is a common pattern in Ricoeur's "Biblical Herme-
> neutics" (1975) and his three-volume *Time and Narrative.* It
> is the pattern in which the function of poetic creation is to
> redescribe the reality which is initially described in the
> prereflective, or ordinary-language, descriptions of reality.
> . . . Initially there is a suspension of reference in the poetic
> creation. This is most evident in purely fictional works. But
> in the metaphorical process the suspension is in the service
> of making a new reference to reality by way of the metaphor-
> ical redescription, showing this reality in a dimension that is
> new when compared with the dimensions shown in the
> ordinary descriptions. This is as true of Ricoeur in the
> three-volume *Time and Narrative* as it is of the earlier
> hermeneutical works.(16)

The point Scharlemann wants to make here is that Ricoeur always
interprets poetic or literary texts as metaphorical redescriptions of
reality. So any poetic activity, including the poetic act of interweaving phenomenological time (fiction) and cosmological time (history), will function as metaphorical redescription. But *re*description
can never constitute an originary presentation of something new or
different. By contrast, the notion of textuality implies that texts are,
or can be, unique individuals. A text can be more than redescriptive:
It can present a world that counterreads the real or existential world

of our original, prereflective descriptions. When a text presents a counterreading in that way, it takes on the presence of a religious symbol in Tillich's sense, as I show later.

Resurrection texts do not simply redescribe the world we already live in, a world that is always already *Sein zum Tode*; they offer something different—a mode of being beyond care and death. To the extent that *Time and Narrative* repeats the pattern of Ricoeur's hermeneutical theory, it falls short of developing a notion of textuality. Texts in their textuality are more than redescriptions of reality. So despite a desire to criticize and move beyond Heidegger's *Sein zum Tode,* Ricoeur's own hermeneutical theory holds him back.

Perhaps now we can appreciate why Scharlemann does not devote attention to Ricoeur's fictive rehabilitation of eternity. Fictive texts, like Mann's *Der Zauberberg,* do not have the power to overturn the existential perspective of *Sein zum Tode* because they only attempt a metaphorical redescription of that perspective. At best, such texts can temporarily suspend the ultimacy of *Sein zum Tode* through imaginative variations. Fictive narratives, complex applications of the metaphorical process, are still in the service of historical consciousness. And, no matter how you slice it, historical consciousness is always limited by ontological death. The difference between Ricoeur's notion of a text as a rediscription of reality and Scharlemann's notion of textuality is analogous to the difference between what Tillich calls a "sign" (which represents an experienced reality) and a "religious symbol" (which gives something new to experience as a matter of ultimate concern). But whereas Tillich was referring to objective things taken as signs or symbols, I refer to the experience of narrated time presented in texts.

Next, Scharlemann attends to the broader pattern in Ricoeur's thought. He notes how Ricoeur develops his notion of narrated time in order to mediate the aporia between phenomenological time and cosmological time. Narrated time, with its attendant sense of historical consciousness, seems to provide a definitive answer to the question "What is time?" But that is not quite the case, as Scharlemann indicates.

> Hence, Ricoeur's own answer to the question "What is time?" is contained in the notion of historical consciousness

as related not just to the distension of the soul but also to real
world history. But that is not the complete answer. The other
part of it, having to do with the inscrutability of time, is
related to the pattern in which Ricoeur, here as in his other
works, moves from the reflective to the hermeneutical or,
more exactly, to the hermeneutics of myths and symbols and
of prereflective rhetoric. (18)

In order to explain the shift in thought that occurs in *Time and
Narrative,* and in prior works by Ricoeur, Scharlemann draws an
analogy between Ricoeur and Kant. He shows that the shift from
reflection to hermeneutics in Ricoeur's thought models the distinction between a rational idea and an aesthetic idea suggested in
Kant's *Critique of Judgement.* "A rational idea," Scharlemann
says, "is defined as a concept for which an adequate intuition can
never be given" (18). The ideals of reason—self, world, and
God—are rational ideas. Each of these ideas contains an insoluble
dialectic of opposite intuitions, which is why they can never be
represented in any one intuition. "An aesthetic idea, by contrast, is
defined as an intuition for which no concept can be adequate"
(19). Aesthetic ideas always give rise to more thought because
concepts can never circumscribe the intuition given in aesthetic
ideas. Ricoeur's understanding of symbols, central to his hermeneutical theory, mimics Kant's understanding of aesthetic ideas.
Scharlemann's analogy, then, runs accordingly:

> A symbol differs from a reflective concept as an aesthetic
> idea does from a rational idea; the one is an intuition in an
> unending search for a concept to interpret it, and the other is
> a concept in a similar search for an intuition to fill it. (19)

An example of this Kantian shift in Ricoeur's thought—from
rational idea to aesthetic idea, or from reflection to hermeneutics—is contained in *Fallible Man.* Reflection can grasp the
possibility of failure or fault on the basis of a theoretical analysis
of human nature, but it cannot retrace the movement from the
possibility of fault to its actuality. Scharlemann notes the repeti-
tion in *Time and Narrative*:

> In the aporetics of time brought out in *Time and Narrative,* a
> similar turn is made, but not until the very end of the work

and as a kind of postscript. In the end it turns out that one cannot say what time is, one can only express indirectly in the language of praise, lamentation, and hope the phenomenon of time. This is, in effect, a turning from reflection on time to another kind of discourse like the interpretation of symbols. (20)

So, in the end, reflection cannot probe the mystery of time at all. It must resign itself to an interpretive role, even a confessional one. Scharlemann does not say so directly, but this shift or resignation is an opening for the new interpretation of Augustine's *Confessions* that Ricoeur proffers in the "Conclusion." He interprets the *Confessions* as a symbolic text that interacts with the mystery of time as it envelops itself within an experience of the eternal. The eternal, in this instance, is the other of time and its mysterious depth.

According to Ricoeur's theory of texts, when discourse is fixed in writing it becomes independent of authorial intention and takes on a partial life of its own. The next step in such a theory, according to Scharlemann, would be to develop a concept of the textuality of texts. Then a text could be viewed in its fullest embodiment as unique individual, capable of *presenting* a world that is not always already a *re*presentation of an existential predicament (as though a more trenchant form of authorial intention survives in the predictable representation of the real in a text's world). Scharlemann develops this "next step" in Ricoeur's hermeneutics in the second and final section of his essay.

Initially, Scharlemann takes a detour through Heidegger's work in order to develop a concept of textuality within the framework of Ricoeur's hermeneutics. This detour makes another analogy, similar to the one made between Ricoeur and Kant; only this time the analogy (between Ricoeur and Heidegger) is implied rather than made explicit. It serves a purpose similar to the earlier Kantian analogy and actually develops it further—toward an understanding of the missing concept of textuality. At issue in both cases is the shift in Ricouer's thought from reflection to hermeneutics, or from one perspective on human nature to another perspective. Scharlemann notes a similar shift in perspective within Heidegger's thought, which is the reason for the conventional distinction between "Heidegger I" and "Heidegger II."[7] He argues that the shift in Heidegger's thought—which in either case is always concerned to answer the same *Seinsfrage* (question of being)—from analysis of Dasein to analysis of artwork and poetic text falls short of developing a notion of textuality even while it points toward such a development. The implied analogy is that Ricoeur's thought falls short of an understanding of textuality, just as Heidegger's does. Scharlemann makes this analogy for the same reason that he made the earlier analogy to Kant: Ricoeur's thought develops out of Heidegger's thought (and Kant's), and thus it is subject to its limitations. Let me be more specific about Scharlemann's analysis of Heidegger.

In *Being and Time,* "Heidegger I" tried to arrive at an understanding of being through an analysis of Dasein as the "there" of being. But he could not make the leap from the careful being in the world of Dasein to the presumably "nonworldly" realm of being itself. This failure to reach being as such led to Heidegger's change of starting point or tactic. As Scharlemann indicates,

> The analysis [of Dasein] which thus culminates in temporal-
> ity as the meaning of being in the world cannot disclose
> whether this temporality is also the horizon for the appear-
> ance of the meaning of being as such. If the being of
> being-in, of indwelling the world, has the form of care, what
> form has being as such? Nothing in the understanding of the
> meaning of being-in makes it possible to leap from there to
> an understanding of being as being. Hence, some beginning
> other than a *Daseinsanalyse* is required. (20)

There are actually two new starting points in "Heidegger II," according to Scharlemann. The first pertains to artwork (*Kunstwerk*), the second to poetic words (*Dichtung*). Only at the level of *Dichtung* is the point of view of being itself made manifest; that is, only the poet says what the philosopher asks. The perspective of *Kunstwerk* stands midway between that of Dasein (showing the meaning of "being-in") and that of *Dichtung* (showing the meaning of being as being). Scharlemann argues that none of the three perspectives of "Heidegger I and II" extends to an understanding of the textuality of texts:

> Textuality is a mode of being other than that of tools that are
> at hand or of objects that are present to us or of other human

beings who are there with us or of an art work or, finally, of
a poetic word. It is, rather, the mode of being which, like
other human beings, is there with us but which, unlike other
human beings, has its place materially defined not by a
physical body but by, let us say, its textual body. To put it in
less Heideggerian language, we can say that in a text we can
not only indwell a world other than that of physical reality
but also encounter another self, the "voice" of the text, just
as we encounter another self in other human existents, and
that this voice of the text need not be identical with the
biographical person who is the author of the text. (15)

After making the analogy between Ricoeur and Heidegger—
the thought of neither one reaches a notion of textuality—
Scharlemann turns his attention to theology and indicates that
Bultmann's concept of kerygma anticipates that of textuality. He
claims that Bultmann's kerygma offers a unique response to the
problems that bible critics have been experiencing with resurrection narratives ever since the publication of Strauss's *Life of
Jesus.* Strauss, like others committed to critical historical understanding, subscribed to the axiom that death is final: Dead people
do not come back to life again. As Scharlemann notes, "The
axiom of the finality of death separates historical from mythical
time-consciousness" (22).[8] Rather than just dismiss the resurrection narratives, Strauss and others tried to explain their origin and
meaning within the context of their mythical form. They tried to
recover a dogmatic interpretation of the resurrection accounts
once criticism had destroyed their empirical value. But, according
to Scharlemann, Bultmann adds something new to this strategy to
reinterpret the resurrection myth. Bultmann claimed that Jesus
was actually resurrected into the kerygmatic text. Scharlemann
states:

> The aim of existentialist interpretation, as Bultmann prac-
> ticed it, goes beyond historical critique, Bultmann's own as
> well as others'. For its intention is to make possible an
> encounter with the same reality of which the disciples spoke
> when they told of the resurrection. It may be true that
> Bultmann, at least now and then, too readily equated the
> mode of being made possible by such a kerygmatic encoun-
> ter with what Heidegger called *Eigentlichkeit,* or authentic-

ity. For there is an important difference between an existence which has death behind it and one which, like authentic existence in Heidegger's analysis, has death ahead of it. But the intention of the concept of kerygma is clear. If Jesus was resurrected into the kerygma, then the kerygma is the materiality of his living presence, and the word of preaching is not concerned with interpreting the meaning of the words of a text but with mediating the living reality in them. (23)

In Scharlemann's interpretation, Bultmann's concept of kerygma indicates the living voice embodied and still heard in the biblical text. It is the "I am" and "Follow me!" of the resurrected Jesus. This is not just Protestant biblicism either, claims Scharlemann. Rather, it indicates the fact that texts have *textuality,* unique voices that individualize texts and mediate a contemporaneousness between them and their readers.[9] Just as the uniqueness of a real person is defined by the here-and-now quality of a body, so too the uniqueness of any text is given in the here-and-now quality of its embodied voice. "In this way," says Scharlemann, "it would be true that the New Testament kerygma is bound with the articulation that is the textuality of New Testament texts" (24).

Once the detour through Bultmann has enabled Scharlemann to articulate a more precise definition of textuality, he returns to Ricoeur's hermeneutics. Scharlemann's conclusion reads as follows:

If a text like Colossians 3:1–3 can read the existence that is Dasein contrary to Dasein's understanding of its own being, if the voice of that writing can be "there too with" us just as much as other human beings, it can do so because its textuality localizes the voice of another way of being in another world just as our psychical and physical bodies localize our existential being. In that other way of being, the temporality of care, which is determined as a temporality in which the self in its futurity—its possibility of having no possibility of being in the world—comes to itself in the actuality of its here-and-now, is converted into the temporality of carefree being (like the lilies of the field that neither toil nor spin). If the being of Dasein is care, and the meaning of that being temporality, then the being that is the textuality of such texts is a freedom from care, and the temporality that is its meaning is one in which the future, namely, the self in its own impossibility [death], has become the past.

Textuality means, in other words, that the literary work, this structured whole of words and sentences, is not only a redescription of the reality of being-there-in-the-world, as it is in Ricoeur's theory of metaphor, but another form of being-there-in-the-world. If this is so, we can ask not only what world is displayed in a work of imagination but also who is the self, who is the voice, that is there in the textual inscription. Ricoeur's hermeneutics does not extend so far as to make such a question possible. Nor does Heidegger's. I have suggested here that the reason lies in the absence of the essential concept of textuality. (25)

The Textuality of the *Confessions*

In "The Textuality of Texts," Scharlemann indicates that the repeated pattern within Ricoeur's thought (from reflection to hermeneutics) always delivers him to the point of religious and theological reflection. But Ricoeur invariably hesitates at that point and never sufficiently develops a religious and theological hermeneutics that can deal with the full range of interpretive problems one encounters in religious texts. Ricoeur's reading of the *Confessions* in *Time and Narrative* follows that same pattern. He begins Part I with a "reflective" analysis of Augustine's philosophical theory of time, and he ends the "Conclusion" with "hermeneutical" questions about the hidden Hebrew archaisms in Augustine's discussion of the Eternal. This change in Ricoeur's treatment of the *Confessions* follows a trajectory that ends with the suggestion of a theological deconstruction. That is, Ricoeur wants to isolate the biblical voice in Augustine's autobiography from its metaphysical trappings. But this deconstruction is never fulfilled. As a philosopher first, Ricoeur reaches only the beginnings of theological reflection.

We can now fulfill Ricoeur's theological deconstruction and demonstrate that the Hebrew sense of the Eternal as divine fidelity that Ricoeur points out is nothing less than the voice of textuality in Augustine's text. This can be accomplished by linking Scharlemann's essay on deconstruction (his employment of Heidegger) with his essay on textuality (his criticism of Heidegger and Ricoeur). The deconstructive essay provides the conceptual tools (specifically, the concept of God's being other) to complete Ricoeur's deconstruction of the *Confessions*. In that context, Ricoeur's discernment of a Hebrew sense of eternity in Augustine's text can be expanded to include the claim that time is the being of God (or the Eternal as divine fidelity) when God is not being God. The essay on textuality, meanwhile, helps to insure that furthering Ricoeur's deconstruction is not simply an arbitrary exercise. It has a specific purpose, which is to illustrate Scharlemann's religious extension of Ricoeur's hermeneutics. Thus, the essay on textuality provides both a control for my furthering Ricoeur's deconstruction and the hermeneutical tools (specifically, the concept of textuality) with which to indicate more precisely the beginnings of this deconstruction of metaphysics within the *Confessions* itself. In that context, the textuality in Augustine's text can be located in the temporally distended soul when it is conceived as the hidden voice of eternity.

Ricoeur's "Conclusion" Fulfilled

In his "Conclusion," Ricoeur confronts the limits of his reflection on time and narrative, which culminates in a brief attempt to interpret the mythic archaisms that persist in philosophical discourse. This confrontation with the inscrutable mystery of time signals the patented shift in Ricoeur's thought from reflection to hermeneutics. In this case, that shift is very late in coming. But come it does. What is significant about it is that it opens the door for religious and theological reflection. The figures of inscrutability that persist in philosophical reflection as mythic archaisms are bound to what is beyond time (eternity, divinity) as it is ordinarily understood. Ricoeur attempts a partial deconstruction of the *Confessions* by insisting that the way in which Augustine conceives the eternal in relation to time is determined by a Hebraic scheme of praise, lamentation, and hope. Ricoeur believes that this Hebrew archaism makes Augustine's thought resistant to the Platonic/metaphysical view wherein eternity only refers to the immutability of a stable present. Instead, he thinks that Augustine's sense of eternity is more deeply indebted to the Hebrew idea of God's fidelity.

If we push this deconstruction to its theological conclusion (which Ricoeur never does himself), three things become apparent: (1) Ricoeur's postcritical understanding of Augustine's eternity as limit-idea can be expanded to include the possibility that the living God is present in time as a manifestation of God's otherness; (2) Augustine's reflections on eternity and time can be expanded to include what he left unthought—namely, how the eternal can be temporally experienced as narrated redemptive time and, as such, can be understood symbolically as the being of God when God is being other than God; and (3) the voice of textuality in the *Confessions* is the voice of eternity as it speaks through the temporally distended soul.

Ricoeur's turn to the mystery of time as manifested in time's other, the eternal, tends to overturn the priorities he established in his previous readings of the *Confessions.* Before the "Conclusion," he spoke of the eternal as an ideal limit for imagination: Eternity is an infinite, hierarchizing aspect of our own being, it is not something "other." But in the "Conclusion," the otherness of eternity comes into consideration. Eternity is said to be a manifestation of *God's* care or fidelity (not *my* care). This new possibility overturns Ricoeur's finite-infinite anthropology as the principal hermeneutic for his confrontation of Heidegger with Augustine. Instead of being the finite's polar opposite, the infinite as other, in this new instance, seeks out the finite and comes into its realm. This infinite is no longer simply a capacity of language or thought within human nature. It is a case of the transcendent manifesting itself in the world. Nor does this new possibility simply repeat an old metaphysical mistake: In precritical metaphysics, the tendency was to identify something finite—the human self or its essential part (the soul)—with the infinite. The soul was said to be immortal, which meant that it naturally belonged to an eternal, transcendent realm. As we now know, this is an instance of transcendental illusion. It takes a tendency of thought—to conceive the self as always already there, hence immortal—too far. Thus, Ricoeur, who is committed to Kantian critical philosophy, must speak of transcendence and eternity within the bounds of human reason and nature if he is to "rehabilitate" such concepts from their metaphysical errancy. But in this new case—the infinite or eternal as other manifesting itself within the finite—the metaphysical tendency is reversed, which means that no subreption has taken place. Nothing that is finite is being called infinite in a transcendent sense (the metaphysical tendency); rather, something truly infinite is manifesting itself as finite.

Nothing within critical philosophy excludes the possibility that the transcendent/eternal can manifest itself in the finite, or even *as* the finite, so long as this manifestation is taken symbolically, that is, as an aesthetic idea. Ricoeur's partial deconstruction of the *Confessions* indicates how a Hebrew conception of eternity and time can sidestep the pitfalls of the traditional theistic picture. The idea of fidelity invites one to think of the Creator as an actor in the created world. The distance between Creator and creation, normally thought to be permanent in its metaphysical interpretation, need not be so overwhelming. The idea of fidelity is an opening for something even further: a reversal in Ricoeur's prior discussion of the eternity/time dialectic. Before the "Conclusion," eternity was said to affect time by serving as limit-idea and as a model for time to imitate. In the "Conclusion," however, this emphasis is potentially reversed. Eternity as fidelity can be said to affect time as a guiding presence from beyond time. If we push Ricoeur's deconstruction a bit further, employing Scharlemann's model, we can say that in certain (religious) instances, time is actually the being of God (as Eternal Fidelity) when God is not being God (that is, when divine being appears as religious symbol). The time that is most affected by eternity, what I have been calling "redemptive time," is a manifestation of eternity in its otherness. So let us seek the textuality of the *Confessions* in Ricoeur's interpretation of them by extending his deconstruction to its theological conclusion.

The objective of deconstructive interpretation, according to Scharlemann, is to think what has been left unthought in a tradition. In the theological tradition, what has been left unthought is the being of the religious symbol, that is, the being of God when God is being other. That means we tend to overlook opportunities to speak of the world as itself a manifestation of God in God's otherness. At least part of the resistance to this kind of conception stems from the metaphysical qualities of our traditional theistic picture. God is pictured as belonging to a transcendent realm far beyond this world of time and change. Yet there are indications of a different picture in the Hebrew archaisms that resist the traditional conception. Fidelity bespeaks a certain worldliness of the divine. What has remained unthought even in the biblical archaisms, however, is how something like the fidelity of God *is* worldly. Fidelity is not just a metaphor for creation, for God's constitution of the world. That would still uphold the theistic picture's view of the distinction between Creator and creation. Ricoeur never challenges this aspect of the theistic picture in his turn to Hebrew fidelity. But if we deconstruct the idea of fidelity further—that is, if we bring Ricoeur's partial deconstruction of the *Confessions* to its theological conclusion—we can begin to see it as a symbol for God in God's otherness (which is closer to the idea of redemption than it is to the idea of creation). Fidelity, in this new light, is God's eternal presence as redemptive time, or as the time that enables the healing of war-torn souls. Time is a war that rages in the soul. Eternity enters into that battle by becoming the time that can turn war into peace. It is as time, as other, that the eternal manifests itself. Even though it is left unthought, Augustine's text in its religious textuality already points to this possibility.

The Eternal Word carries the sense of fidelity in the *Confessions* because it is the Word that prompts Augustine to turn away from sin and guides him back to God. As Augustine says,

> He [the Word, or Christ] is therefore the Beginning, the abiding Principle, for unless he remained when we wandered in error, there would be none to whom we could return and restore ourselves. But when we return from error, we return by knowing the Truth; and in order that we may know the truth, he teaches us, because he is the Beginning and he also speaks to us [that is, he is also the Ending]. (Book XI, ch. 8, 260)

As Eternal Word, God is solicitous about the welfare of humans. That is how fidelity characterizes eternity in the *Confessions.* How is this solicitude expressed? The Word, says Augustine, "speaks" from within the soul. I understand this to mean that the Word speaks as narrated time, a time that Augustine himself narrates as a time of redemption. Since the soul is a temporal phenomenon, it speaks by prompting memories and expectations: It distends the soul as it orchestrates salvation. By its presence, which "fills us to the brim," the Word even enables us to forget our sins.[10]

So God is more than just the source of a "physical" time: God is the *agency* of time. Time, like everything else, has its existence in God (Book I, ch. 6, 27). For Augustine, that everything exists *in* God means that God surrounds it in its temporal extension. As Beginning, God extends to the "before" of time's creation; and as Ending, God extends to the "after" of time's outcome. God's eternal life includes the timing of time as agent of creation (Beginning) and redemption (Ending). Because God is (as eternal), our being (as temporal) is possible. Yet our being in the world is always restless. Feeling trapped in the finite, we strive to become more like the infinite that surrounds us and sustains us. Time imitates eternity, claims Augustine. It strives to become like the eternal peace that comforts it even in its most chaotic distensions.

On practically every page of the *Confessions* one can find references to God's agency in the world. God's solicitude extends everywhere. No one can accuse Augustine of leaving unthought how God, as Eternal Word, is in the world. But what he does leave unthought is how, with regard to the human soul, God *is* the world. Augustine points to the religious experience of hearing the Word in one's soul. But he does not think through all the implications of this experience. The question is: What is the nature of God's agency in time? What distinguishes redemptive time (God acting in the soul) from creative time (God acting in the world)? Augustine does not adequately distinguish these two different senses of time.

The problem is that Augustine interprets God's acting in the soul (redemptive time) as an instance of God's acting in the world (creative time). And since he is more concerned with the former than with the latter, it appears that Augustine actually collapses the latter into the former. This is what enabled Ricoeur to claim that Augustine only offers a psychological interpretation of time. But that is not entirely true. It is true, however, that he is more interested in "psychological" (or redemptive) time than in cosmological (or creative) time. But God's acting in the soul as redemptive time is more than just another instance of God's solicitous presence everywhere in the world. God's solicitous presence in the world at large can be understood within the theistic picture, but God's solicitous presence within the soul itself cannot. In the latter case, God's temporal agency—God's willingness to act in the world—is transformed into *my* being. God exercises the agency of the time of the soul so that it can turn away from sin and turn toward its final end. Redemptive time, the time of the soul in its striving to become eternal, is the being of God (as Eternal Fidelity) when God is not being God. It seems to me that this is precisely what Augustine means (though it remains unthought) when he says things such as "God is closer to my being than I am" or "God is the life of my life."[11]

The metaphysical constraints of the theistic picture prevent Augustine from completely thinking through this distinction between God's acting in the world and God's becoming the temporal soul as an expression of God's otherness. But with some deconstructive help, we can see how Augustine himself could think that the final outcome of divine fidelity is for God to take on the life of the other, to become time in order to redeem it. But Augustine could not quite conceive an eternity that gives up its eternality for the sake of time—though he could experience it and allude to it as the teaching office of the Word. His sense of eternity is still locked within the metaphysical definition. It would not make sense for eternity, as always present and never distended, to become like time. As Ricoeur indicates in his "Conclusion," there is a tension in Augustine's text between eternity understood as fidelity and eternity understood as presence. The latter sense is stronger, but it is possible for us to think through the other sense for Augustine now that the metaphysical cast of the theistic picture has been broken. When we do that, we hear again the silent divine voice that speaks through Augustine's own confession as its textuality.

According to Scharlemann, the key feature in the religious textuality of a text is entailed in the text's ability to put the care and death of Dasein's existential being behind it:

> If the being of Dasein is care, and the meaning of that being temporality, then the being that is the textuality of such texts is a freedom from care, and the temporality that is its meaning is one in which the future, namely, the self in its own impossibility [that is, its death], has become the past. ("The Textuality of Texts," 23)

For Ricoeur (who employs Heideggerian categories in his hermeneutics), texts represent possibilities in the existential world

that always have death before them. But there are cases when texts
present a world that has death behind it. Religious texts often do
that, and it is a voice of textuality that makes their transgression of
existential limits possible.

Scharlemann's extension of Ricoeur's theory of text to include a
notion of textuality can be applied to Ricoeur's interpretation of
Augustine's text. In fact, the *Confessions* is able to verify Scharlemann's notion of textuality by speaking of the precise situation that
demarcates the difference between a text whose textuality or
religious voice is not made manifest (a text with death before it) and
a text whose textuality is made manifest (a text with death behind it).
In one of his prayers, Augustine says this: "For all I want to tell you,
Lord, is that I do not know where I came from when I was born into
this life which leads to death—or should I say, this death which
leads to life?" (Book I, ch. 6, 25). What he is suggesting is that once
the soul becomes aware of the living Word inside itself, once the
chaotic distension of the soul is recognized as sin, death is no longer
of great consequence. Within the life of repentance, death no longer
stands in front of the self but behind it: To turn away from sin is
already to have "died." The voice of textuality that announces this
new circumstance in the world is the voice of faith (as God's own
voice) speaking through the confessions of a contrite soul. "It is my
faith that calls to you, Lord, the faith which you gave me and made
to live in me through the merits of your Son, who became man, and
through the ministry of your preacher" (Book I, ch. 1, 21).

A New Confrontation Between Heidegger and Augustine

By following through the trajectory of Ricoeur's interpretation
of the *Confessions,* and by extending his partial deconstruction of
Augustine's sense of eternity, we can see how the entirety of the
Confessions is bound up within the dialectic of time and eternity.
This dialectic provides the basis for an argument that the two parts
of the *Confessions* form a coherent unity. What is narrated in the
first nine or ten books is grounded by the theological reflections
inspired by the Bible that appear in Augustine's last three or four
books. The discussion of time and eternity in Book XI is the
fulcrum of the entire text. Time not only imitates eternity; it is the
Eternal when the Eternal is not being the Eternal. Redemptive
time is the actual presence of God in the temporal soul. This time,
as God's being in its otherness, prompts the soul to confess its sins

and return to God. The soul itself, as time, becomes the mysterious agency of God's saving grace. Thus, the whole process of redemption, which includes the *confessio,* is locked within the mystery of time. Ricoeur's interpretation of Augustine's theory of time, when pushed in a direction that he himself suggests, provides an excellent argument for the unity of the *Confessions.*[12] What actually unites the text is the constancy of its own textuality, which speaks an Eternal Word through human words. This voice is as apparent in Augustine's memories of growing up as it is in his exegesis of Genesis 1:1.

Despite great historical and cultural distance, I think we can still hear the religious voice of textuality that speaks through Augustine's text. The eternal that is an aspect of God's being, and not our own, is still discernible in its otherness. Criticism may have debunked certain metaphysical conceptions of a transcendent realm, but not the possibility that the transcendent *is* the finite in the case of religious experience. After extending Ricoeur's theory of text and his reading of the *Confessions,* it is possible to speak of a new confrontation between Heidegger and Augustine. The eternal in its otherness, in its ability to become my own temporal being, is not confinable to the existential world of historical consciousness. It speaks of a world that is beyond *Sein zum Tode.* Because we can have the experience of faith that calls forth such a mode of being, we can understand religious texts such as the *Confessions* that project such a world where death is in the past. Heidegger's text analyzes the existential world where death always lies ahead. Augustine's text manifests a faith world where death lies behind. So long as we can still hear and understand the voice of textuality in Augustine's *Confessions,* the two texts need not be in conflict at all.

A Textual Correlation

To support the claim that the *Confessions* and *Being and Time* are not conflicting texts, let me return to the post-Kantian theological context of Bultmann and Tillich within which my argument is made. The preached kerygma enables a religious "answer" to the problem of death for Bultmann. The Christian symbol of Eternal Life enables such an answer for Tillich. In Augustine's *Confessions,* according to Ricoeur's reading as I have extended it, the voice of textuality in narrating the time of redemption enables a reader—any reader—to put death behind and attain the path of redemption. The point here is that the time in which God enters the soul is the *narrated* time of redemption for Augustine. Augustine's own accounts of time and eternity point beyond metaphysical conceptions to the *experience* of God's eternity *through* narrated time *as* religious symbol. Consider the way in which Augustine himself narrates the experience of redemptive time in the *Confessions*: He narrates the impact of his own *reading* of a text with the power of textuality.

In the famous *tolle lege* scene (Book VII, ch. 12), Augustine's spiritual crisis comes to a head. He has reached a point of despair concerning his own inability to master the conflict of wills or desires—for God or for worldly pleasures—that rages in his soul. With Alypius, he retires to the garden of their house in Milan. Augustine weeps under a fig tree. Suddenly he hears a child's voice singing:

> Whether it was the voice of a boy or a girl I cannot say, but again and again it repeated the refrain "Take it and read, take it and read" ["*tolle lege, tolle lege*"]. At this I looked up, thinking hard whether there was any kind of game in which children used to chant words like these, but I could not remember ever hearing them before. I stemmed my flood of tears and stood up, telling myself that this could only be a divine command to open my book of Scripture and read the first passage on which my eyes should fall. (Book VIII, ch. 12, 177)

Augustine took the advice of the child's voice and opened his bible at random. The passage he turned to contained Paul's letter to the Romans. In part, it said, "Arm yourselves with the Lord Jesus Christ; spend no more thought on nature and nature's appetites" (Rom. 13:13). Upon reading that text in the garden, according to Augustine, his life was miraculously transformed. He was now, in heart and in will, a converted Christian.

What is interesting about this conversion is the role that a text plays. Augustine turns to the biblical text in order to find a personal, divine message that will set him free from sin. He is not disappointed. How are we to understand this seemingly arbitrary exercise? I suggest that what is at work here is the voice of textuality within the biblical text.

In reading Paul's text, Augustine experiences the Eternal

Verbum as it becomes the redemptive time of his own life. The biblical voice of textuality in Paul's letter speaks through Augustine's own text in the *Confessions.* The *Confessions* themselves narrate the experience of responding to the textuality of Paul's letter to the Romans. The biblical text works within the *Confessions* as Augustine wrote them just as the *Confessions* themselves work within our lives as we read them. Through the symbolic power of language and the narrated time of its interpretation, the textuality disclosed in the biblical text for Augustine is now available for us in Augustine's own text. That is the case because it was Augustine's encounter with the Eternal in the garden of Milan, through Paul's text, that enabled him to make sense of the disparate forces in his own life and to narrate them in his *Confessions.* The poles of his own narrative—the philosophical texts of the Platonists that helped to free him from the material dualism of the Manichees; the biblical texts (at first unappealing to the rhetorician) that were made more credible by the kerygmatic preaching of Ambrose—are suddenly brought together or correlated because of the textual event in the garden. There is, then, a reflexive quality to the textuality of the *Confessions.* It repeats for us, in Augustine's own life, the textuality of Paul's epistle. It embodies the textuality—the Eternal *Verbum* as redemptive voice and redemptive time—of which it speaks (and to which it points) for our benefit as readers.

Narrated redemptive time functions as the presence of God when God is not being God. In this textual experience, the criteria of a religious symbol are met. Narrated redemptive time in the *Confessions* does the following: (1) It binds subject and object together in ultimate concern, and (2) it denies that it is *literally* God (since it is only the time that *points to* and *promises* eternity). In these two factors taken together, the promise of a redemptive presence is fulfilled.

Conclusions

A key message in Ricoeurean hermeneutics has always been this: Let us not be afraid to criticize the tyranny of criticism itself. That is the way to continue the project of criticism beyond its own dogmatic tendencies. That is the way, says Ricoeur, to hear again the ancient call of religious symbols and belief. Heidegger's notion of *Sein zum Tode,* which depicts an ontological interpretation of death as ultimate temporal horizon, is a critical dogma that is in need of more criticism. Ricoeur criticizes Heidegger on death from a philosophical perspective that, in the end, engages a theological and religious perspective. He confronts (or correlates) Heidegger on time and death with Augustine on time and eternity. This study extends Ricoeur's argument theologically such that a new confrontation between Heidegger and Augustine can appear. This new confrontation enables us to bring Ricoeur's hermeneutical project all the way home—to the point where the Eternal *Verbum* can once again be heard and understood as it speaks to us from within the textuality of Augustine's ancient text. The religious world that it calls us to enter in faith has already put death behind it.

Once extended with the help of Scharlemann's theological criticisms, Ricoeur's hermeneutics enable us to read religious classics such as the *Confessions* with renewed insight. Narrated time, in its mysterious depth, is a manifestation of God's redemptive being, according to Augustine. In light of the experience of redemptive time through textuality, death is of little consequence. Instead of being-toward-death, our days become part of a foretaste, an anticipation, of an eternal life to come. Neither Bultmann nor Tillich nor Scharlemann—all theologians influenced by Heidegger—gave up a sense of the eternal because of Heidegger's philosophy. Following Ricoeur and Scharlemann, I have demonstrated the wisdom of such tenacity. It is possible for the modern critical reader to encounter the divine in the textuality of the *Confessions*: Our reading of Augustine's encounter with textuality points, like a symbol, toward the possibility of a similar encounter for us. To put it differently, the *noesis* that is disclosed in Augustine's narration becomes the modern reader's *noema.*

Notes

1. See the next section, "Beginning a Theological Extension," where I shall make this connection more explicit. As we shall see, Scharlemann argues that Ricoeur repeats a strategy from his earlier writings in the "Conclusion" to *Time and Narrative*. Earlier, Ricoeur initiated a methodological shift—from theoretical reflection to hermeneutical interpretation—in order to accommodate the difference between an inquiry into the possibility of human evil (*Fallible Man*) and an inquiry into the fact of it (*Symbolism of Evil*). Just as we must revert to a symbolics of evil in order to speak of the fact of evil, we must revert to a mythics of time in order to speak of its primordial facticity, or of its already having enveloped us. Hence Ricoeur's claim that time is inscrutable, as is evil. Time precedes us and envelops us, as does evil.

2. See chapter 2 herein. Also note that Ricoeur finally concedes my earlier point about the theological context of Augustine's reflections on time. Augustine praises the Creator, and his hope is in the Redeemer; what he laments is that the two works of the Eternal Word do not coincide in his distended soul. The discrepancy between praise and hope, or between creation and redemption, actually bespeaks two different senses of time. I mention this in order to emphasize that Ricoeur is offering a new interpretation of the *Confessions* in his "Conclusion," one that is more theologically motivated.

3. "The difference in levels between Augustine's thought and the Hebraic thinking, which constitutes his archaism, is concealed by the Greek, and then by the Latin, translation of the well-known *ehyeh asher ehyeh* in Exodus 3:14a. The Revised Standard Version of the Bible has 'I am who I am,' as do current French translations. But thanks to this ontologizing of the Hebraic message, we occlude all the senses of eternity that rebel against Hellenization. For example, we thereby lose the most precious sense, whose best equivalent in modern language is expressed by the idea of fidelity. The eternity of Jahweh is above all else the fidelity of the God of the covenant, accompanying the history of his people" (III, 265).

4. Scharlemann derives this definition of text from Ricoeur's "Qu'est-ce qu'un texte?" See "The Textuality of Texts," page 14, for bibliographic information.

5. Scharlemann focuses on St. Paul's letter to the Colossians: "If then you have been raised with Christ, seek the things that are above, where Christ is, seated at the right hand of God. Set your minds on things that are above, not on things that are on earth. For you have died, and your life is hid with Christ in God. When Christ who is our life appears, then you also will appear with him in glory" (Col. 3:1–2). Scharlemann comments that "such a being beyond death fits neither with existential time nor the time of historical consciousness, nor does this excerpt from a Pauline letter fit into any of the genres that Ricoeur discusses in his biblical hermeneutics, which culminates in the theologicial mediation of religious expressions" (15).

6. This shift or movement is entailed in the two parts of volume 2 of *Philosophy of the Will*: In *Fallible Man,* Ricoeur inquires into the conditions of possibility for evil or fault; in *The Symbolism of Evil,* he interprets the confessional symbols that indicate the fact of evil or fault.

7. This convention was introduced by William Richardson, in *Heidegger: Through Phenomenology to Thought.*

8. Note the implications of this axiom for my claim that death has increased in its ontological significance since the advent of a critical perspective. Coincident with the rise of critical thought is an equally significant rise in historical consciousness. The two are not unrelated, though their exact relationship would be difficult to pin down.

9. Scharlemann does not make a connection to Kierkegaard here, but I think such a connection is appropriate. Hence my use of the term "contemporaneousness." See chapters 4 and 5 of *Philosophical Fragments,* where Kierkegaard discusses the problem of contemporaneity in Christian discipleship. He takes up the issue in response to a remark that Lessing made, namely, that it would be easier to be Christian if one lived at the time of Christ. At issue is how the eternal can be operative within historical times, whether in the first century or the nineteenth. This is a key interpretation of the paradox that so delights Kierkegaard.

10. "To whom shall I turn for the gift of your coming into my heart and filling it to the brim, so that I may forget all the wrong I have done and embrace you alone, my only source of good?" (Augustine, *Confessions,* Book I, ch. 5, 24)

11. "But you are the life of souls, the life of lives" (*Confessions,* Book

III, ch. 6, 62). "He [God] is the Life of the life of my soul" (Book X, ch. 6, 213).

12. One of the things that differentiates my Ricoeur-inspired deconstruction of the *Confessions* from the deconstructions made by Joseph O'Leary (*Questioning Back: The Overcoming of Metaphysics in Christian Tradition*) and Mark Taylor (*Erring: A Postmodern A/theology*) is my concern to place such an interpretation within the broader context of *Confessions* scholarship over the past century. Taylor in particular seems indifferent to the efforts of others to offer both critical and postcritical readings of the *Confessions.*

Bibliography

Altizer, Thomas J. J., ed. *Deconstruction and Theology*. New York: Crossroad Publishing Co., 1982.

Arendt, Hannah. *The Human Condition*. Chicago, IL: The University of Chicago Press, 1958.

————. *The Origins of Totalitarianism*. New York: Harcourt Brace Jovanovich, 1973.

Aristotle. *Physics*. Translated by R. P. Hardie and R. K. Gaye. In *Great Books of the Western World*, ed. R. Hutchins. Volume 8, *The Works of Aristotle: Volume I* (259–355). Chicago, IL: Encyclopaedia Britannica, Inc., 1952.

————. *The Rhetoric and the Poetics of Aristotle*. Edited by F. Solmsen. New York: Random House, 1954.

Augustine. *Confessions*. Translated by R. S. Pine-Coffin. New York: Penguin Books, 1961.

————. *Confessionum*. Edited by M. Skutella. Stuttgart: B. G. Teubner, 1981.

Becker, Ernest. *The Denial of Death*. New York: Free Press, 1973.

Brown, Peter. *Augustine of Hippo*. Berkeley: University of California Press, 1967.

Bultmann, Rudolph. "The Historicity of Man and Faith." In *Existence and Faith*, edited by S. Ogden (92–110). New York: Harper and Row, 1962.

————. "New Testament and Mythology." In *New Testament and Mythology and Other Basic Writings*, edited and translated by S. Ogden (1–44). Philadelphia, PA: Fortress Press, 1984.

Cassirer, Ernst. "Kant and the Problem of Metaphysics: Remarks on Martin Heidegger's Interpretation of Kant." In *Kant: Disputed Questions*, edited by M. S. Gramm (167–93). 2d ed. Atascadero, CA: Ridgeview Publishing Co., 1984.

Gadamer, Hans-Georg. *Truth and Method*. Translated by G. Barden and J. Cumming. New York: Continuum Publishing Corp., 1975.

————. "Martin Heidegger and Marburg Theology." In *Philosophical Hermeneutics*, edited and translated by D. Linge (198–212). Berkeley: University of California Press, 1976.

Guitton, Jean. *Le Temps et l'Eternite chez Plotin et saint Augustin*. Paris: Vrin, 1933.

Hegel, Georg Wilhelm. *Lectures on the Philosophy of Religion,* vol. 3. Edited by P. Hodgson. Berkeley: University of California Press, 1985.

Heidegger, Martin. *Being and Time.* Translated by J. Macquarrie and E. Robinson. New York: Harper and Row, 1962.

———. *Kant and the Problem of Metaphysics.* Translated by J. Churchill. Bloomington: Indiana University Press, 1962.

———. "Language." In *Poetry, Language, Thought,* translated by A. Hofstadter (187–210). New York: Harper and Row, 1975.

———. *Sein und Zeit.* Vol. 2 of *Gesamtausgabe.* Frankfurt am Main: Vittorio Klostermann, 1977.

Hick, John. *Death and Eternal Life.* New York: Harper and Row, Publishers, 1976.

Husserl, Edmund. *The Phenomenology of Internal Time-Consciousness.* Edited by M. Heidegger. Translated by J. Churchill. Bloomington, IN: Indiana University Press, 1964.

———. *Ideas Pertaining to a Pure Phenomenology and to a Phenomenological Philosophy,* first book. Translated by F. Kersten. Boston, MA: Martinus Nijhoff Publishers, 1983.

Jaspers, Karl. *The Atom Bomb and the Future of Man.* Translated by E. B. Ashton. Chicago, IL: The University of Chicago Press, 1961.

Kant, Immanuel. *Critique of Judgement.* Translated by J. Meredith. In *Britannica: Great Books,* vol. 42, edited by R. Hutchins (461–613). Chicago, IL: Encyclopaedia Britannica, 1952.

———. *Religion within the Limits of Reason Alone.* Translated by T. Greene and H. Hudson. New York: Harper and Row, 1960.

———. *Critique of Pure Reason.* Translated by N. K. Smith. Unabridged ed. New York: St. Martin's Press, 1965.

———. *Critique of Practical Reason.* Translated by L. Beck. New York: Macmillan Publishing Co., 1985.

Kierkegaard, Søren. *Philosophical Fragments.* Translated by D. Swanson and H. Hong. Princeton, NJ: Princeton University Press, 1967.

Klemm, David. *The Hermeneutical Theory of Paul Ricoeur: A Constructive Analysis.* Lewisburg, PA: Bucknell University Press, 1983.

———. *Hermeneutical Inquiry, Volume I: The Interpretation of Texts.* Atlanta, GA: Scholars Press, 1986.

———. *Hermeneutical Inquiry, Volume II: The Interpretation of Existence.* Atlanta, GA: Scholars Press, 1986.

Kübler-Ross, Elisabeth. *On Death and Dying.* New York: Macmillan Publishing Co., 1969.

Meijering, E. P. *Augustin uber Schopfung, Ewigkeit und Zeit: Das elfte Buch des Bekenntnisse.* Leiden: E. J. Brill, 1979.

Moltmann, Jurgen. *Theology of Hope.* Translated by J. Leitch. New York: Harper and Row, 1967.

O'Leary, Joseph. *Questioning Back: The Overcoming of Metaphysics in Christian Tradition.* Minneapolis, MN: Winston Press, 1985.

O'Meara, Thomas F. "Tillich and Heidegger: A Structural Relationship," *Harvard Theological Review* 61 (1968): 249–61.

Plato. *Plato: The Collected Dialogues.* Edited by E. Hamilton and H. Cairns. New York: Bolingen Foundation, 1961.

Pöggeler, Otto. *Martin Heidegger's Path of Thinking.* Translated by D. Magurshak and S. Barber. Atlantic Highlands, NJ: Humanities Press International, Inc., 1987.

Reagan, Charles. "Review of *Time and Narrative,*" *International Philosophical Quarterly* 25 (1985): 89–105.

Richardson, William. *Heidegger: Through Phenomenology to Thought.* The Hague: Martinus Nijhoff Publishers, 1963.

Ricoeur, Paul. *The Symbolism of Evil.* Translated by E. Buchanon. Boston, MA: Beacon Press, 1967.

———. "Biblical Hermeneutics," *Semeia: An Experimental Journal for Biblical Criticism* 4 (1975): 27–148.

———. *Interpretation Theory: Discourse and the Surplus of Meaning.* Fort Worth: Texas Christian University Press, 1976.

———. *The Rule of Metaphor.* Translated by R. Czerny. Toronto: University of Toronto Press, 1977.

———. "The Function of Fiction in Shaping Reality," *Man and World* 12 (1979): 123–41.

———. "Preface to Bultmann." In *Essays on Biblical Interpretation,* edited by L. Mudge (49–72). Philadelphia, PA: Fortress Press, 1980.

———. *Temps et recit,* vols. 1 and 3. Paris: Editions du Seuil, 1983.

———. *Time and Narrative,* vols. 1 and 2. Translated by K. McLaughlin and D. Pellauer. Chicago, IL: The University of Chicago Press, 1984, 1985.

———. *Fallible Man.* Translated by C. Kelbley. Rev. ed. New York: Fordham University Press, 1986.

———. "On Interpretation." In *After Philosophy: End or Transformation?,* edited by K. Baynes, J. Bohman, and T. McCarthy (357–80). Cambridge, MA: MIT Press, 1987.

———. *Time and Narrative,* vol. 3. Translated by K. Blamey and D. Pellauer. Chicago, IL: The University of Chicago Press, 1988.

Scharlemann, Robert. *The Being of God.* New York: Seabury Press, 1981.

———. "The Being of God When God Is Not Being God." In *Deconstruction and Theology,* edited by T. J. J. Altizer (79–108). New York: Crossroad Publishing Co., 1982.

———. "Being Open and Thinking Theologically." In *Hermeneutical*

Inquiry, Volume II: The Interpretation of Existence, edited by D.
Klemm (259–71). Atlanta, GA: Scholars Press, 1986.

―――. "The Textuality of Texts." In *Meanings in Texts and Actions:
Questioning Paul Ricoeur,* edited by D. Klemm and W. Schweiker
(13–25). Charlottsville: University Press of Virginia, 1993.

Schleiermacher, Friedrich. *Soliloquies.* Translated by H. Friess. Chicago, IL: Open Court Publishing Co., 1926.

―――. *Hermeneutics: The Handwritten Manuscripts.* Translated by J.
Duke and J. Forstman. Atlanta, GA: Scholars Press, 1977.

Schweiker, William. "Beyond Imitation: Mimetic Praxis in Gadamer,
Ricoeur, and Derrida." *Journal of Religion* 68 (January 1988): 21–38.

Shannon, Thomas, ed. *Bioethics.* 3d ed. Mahwah, NJ: Paulist Press,
1987.

Taylor, Mark. *Erring: A Postmodern A/theology.* Chicago, IL: The
University of Chicago Press, 1984.

Tillich, Paul. *The Interpretation of History.* New York: Scribner's, 1936.

―――. *Systematic Theology.* 3 vols. Chicago, IL: The University of
Chicago Press, 1951, 1957, 1963.

―――. *The Courage to Be.* New Haven, CT: Yale University Press,
1952.

―――. *The Theology of Paul Tillich.* Edited by C. Kegley and R.
Brettal. New York: Macmillan Publishing Co., 1952.

―――. *Dynamics of Faith.* New York: Harper and Row, 1957.

―――. "The Word of God." In *Language: An Enquiry into Its Meaning
and Function,* edited by R. Anshen (160–209). New York: Harper and
Brothers, 1957.

―――. *Theology of Culture.* Edited by R. Kimball. New York: Oxford
University Press, 1959.

Walhout, Clarence. "On Symbolic Meanings: Augustine and Ricoeur."
Renassence, vol. 31, no. 2 (Winter 1979): 115–27.

Weir, Robert, ed. *Ethical Issues in Death and Dying.* New York:
Columbia University Press, 1977.

Welch, Claude. *Protestant Thought in the Nineteenth Century.* 2 vols.
New Haven, CT: Yale University Press, 1972, 1985.

Wordsworth, William. *The Prelude: 1799, 1805, 1850.* Edited by J.
Wordsworth, M. H. Abrams, and S. Gill. New York: W. W. Norton
and Co., 1979.

Wyschogrod, Edith. *Spirit in Ashes: Hegel, Heidegger, and Man-Made
Mass Death.* New Haven, CT: Yale University Press, 1985.

―――, ed. *The Phenomenon of Death: Faces of Mortality.* New York:
Harper and Row, 1973.

Index

Index

Pascal, Blaise 29
Paul, Saint 11, 27, 154–155, 158
Phenomenological/psychological time 15–16, 19–21, 29, 31–32, 52, 56–63, 68, 71, 73, 75–77, 88, 92, 106, 109–111, 131, 133, 138–139, 150
Plantinga, Alvin 12
Platonic forms 10–11, 27
Postcritical idea of God 2–4
Post-Kantian theology 2, 12, 88, 153
Proust, Marcel 62

Quinlan, Karen Ann 28

Redemptive time 14, 16, 22, 32, 54, 119, 125, 147–148, 150–152, 155–156
Religious symbol 3–4, 6–8, 12–14, 21, 66, 91, 117–118, 120–122, 124–125, 132, 138–141, 147–149, 154–158
Richardson, William 158, 163
Ricoeur, Paul 12, 15–23, 27, 29, 31–36, 38–40, 42–46, 48, 50, 52–68, 71–76, 81–83, 86–89, 91–92, 95, 106–111, 121–122, 128, 131–149, 151–153, 155, 157–159, 163–164
Russell, Bertrand 27

Scharlemann, Robert 2–3, 6–7, 12, 16–17, 22–23, 26–27, 67, 122–124, 129, 132, 136–146, 148, 151–152, 156–158, 164
Schleiermacher, Friedrich 13, 29, 129, 164
Sein zum Tode 1, 8, 11, 15, 20–21, 26–27, 29, 95, 101, 103, 107–111, 114–116, 119–122, 129, 131, 137, 139, 153, 156

Self 2, 8–9, 21, 28, 76, 84–85, 87, 96–97, 99–100, 103, 105, 109, 111, 119, 121, 126, 137, 139–140, 143–145, 147
Socrates 26
Soul 8–9, 12, 15, 29, 32–34, 43–47, 51–53, 55, 59–60, 63, 67, 70–77, 80, 88, 117, 125, 135, 140, 146–147, 149–154, 157–159
Strauss, David 143

Taylor, Mark 159, 164
Temporality 8–9, 11, 14, 32, 38, 53, 58, 60, 72, 81, 96–97, 102, 106–110, 112, 120–121, 132, 142, 144, 151
Textuality 16–17, 23, 67, 132, 136–139, 141–149, 151–156, 158
Theology 1–3, 9, 11–14, 16–18, 22, 28, 52, 54, 56, 63–64, 67, 105–106, 112–119, 121–124, 128–129, 131–132, 134–137, 143, 145–146, 148–149, 152, 156–158
Tillich, Paul 2–3, 6–7, 12–14, 22, 26, 113, 116–122, 129, 139, 153, 156, 164
Time 8–11, 14–21, 23, 31–36, 38–63, 67–89, 92–93, 95–97, 101–103, 105–111, 118–125, 131–137, 139–141, 143, 145–154, 156, 158
Tolle lege 154
Transcendence 8–12, 21, 26–27, 29, 62, 77, 82–84, 93, 103–104, 106, 111–112, 117, 119, 122, 147–148, 153
Transcendental illusion 9, 106, 111, 147

Woolf, Virginia 62
Wordsworth, William 29, 164
Wyschograd, Edith 25, 164

About the Author

Richard Severson is a graduate of South Dakota State University. He earned his Ph.D. from the University of Iowa. Dr. Severson taught as a visiting professor at the University of Northern Iowa. He has been on the library faculty at George Fox College and Linfield College, and currently works at the public library in Tigard, Oregon. Dr. Severson is a member of the American Philosophical Association and the American Academy of Religion.